AF566747

VACUUM MICROBALANCE TECHNIQUES

VOLUME 7

VACUUM MICROBALANCE TECHNIQUES

Volume 1
Fort Monmouth Conference–1960
Edited by Max J. Katz

Volume 2
Washington, D.C. Conference–1961
Edited by Raymond F. Walker

Volume 3
Los Angeles Conference–1962
Edited by Klaus H. Behrndt

Volume 4
Pittsburgh Conference–1964
Edited by Paul M. Waters

Volume 5
Princeton Conference–1965
Edited by Klaus H. Behrndt

Volume 6
Newport Beach Conference–1966
Edited by A. W. Czanderna

Volume 7
Eindhoven Conference–1968
Edited by C. H. Massen
and H. J. van Beckum

Volume 8
Wakefield Conference–1969
Edited by A. W. Czanderna

VACUUM MICROBALANCE TECHNIQUES

VOLUME 7

Proceedings of the Eindhoven Conference
June 17-18, 1968

Edited by
C. H. Massen and H. J. van Beckum

Eindhoven University of Technology
Eindhoven, Netherlands

PLENUM PRESS • NEW YORK–LONDON • 1970

Library of Congress Catalog Card Number 61-8595

SBN 306-38407-8

A Division of Plenum Publishing Corporation
227 West 17th Street, New York, N.Y. 10011

United Kingdom edition published by Plenum Press, London
A Division of Plenum Publishing Corporation, Ltd.
Donington House, 30 Norfolk Street London W.C. 2, England

Printed in the United States of America

At the Seventh Vacuum Microbalance Techniques Conference one member was unfortunately absent whom all his colleagues would have been glad to meet again. He was our friend Lee Cahn, the initiator and driving force of these conferences, whose sudden death deprived the field of microbalance research of a most enthusiastic collaborator.

Cahn's inventive faculty has always inspired young research workers all over the world and has led to the extension of our knowledge, the improvement of vacuum microbalance techniques, and the expansion of their fields of application.

Lee Cahn will be especially remembered for his active part in the conference discussions, which resulted in many novel views. It was Cahn's practical mind that turned them into improvements of his products. In fact, Lee Cahn was a link between science and engineering, and he will be sorely missed.

Editors' Note

The Seventh Vacuum Microbalance Techniques Conference at the Eindhoven University of Technology, Netherlands, was the first to be held outside the U.S.A. The great number of participants and the lively discussions reflected the broad interest in this field in Europe. The Conference also demonstrated that the development of techniques persists in being the subject of many varied investigations.

Of the 23 papers presented at the Conference 21 have been incorporated in the present volume in alphabetical order of first authors' names. Contributions include descriptions of developments of unconventional weighing techniques (Gast and Poulis et al.) and papers dealing with the simultaneous use of different weighing systems (Hillecke et al.). The techniques of calibration (Schmider et al.) and the different techniques used to eliminate weighing errors (Cutting, Kuhn et al., Moret et al., Pebler, and Robens et al.) are reported together with experimental data and theoretical discussions.

The multifarious applications of vacuum microbalances in scientific research are shown in papers concerning such fields as magnetic phenomena (Van den Bosch, Bransky et al., and Cini et al.) and chemical reactions (Boudeulle et al., Cameron et al., Clough et al., Dovaston et al., and Gulbransen et al.) under widely diverging experimental conditions, e.g., temperature and gas pressure.

Other contributions – more or less of a synoptical nature – deal with vacuum system evaluation and the use of vacuum microbalances (Kollen et al., Gregg, and Wiedemann), one paper (Dijkema et al.) describing the microgravimetrical determination of diffusion coefficients in vapors.

The editors are very grateful to Mrs. F. Duifhuis-Van Tongeren for her secretarial work and to Miss M. Gruyters for her share in the preparation of the graphic work.

C. H. Massen

H. J. van Beckum

Contents

Conference Participants

A. Adam: Institute de Physique, Nancy, France.

J. M. P. C. van Asten: Eindhoven University of Technology, Eindhoven, Netherlands.

G. Bapst: Laboratoire de Minéralogie, Strasbourg, France.

H. J. van Beckum: Eindhoven University of Technology, Eindhoven, Netherlands.

A. Van den Bosch: F. V. S. –S. C. K., Mol, Belgium.

M. Boudeulle: Laboratoire de Minéralogie, Lyon, France.

I. Bransky: Technion Technical Institute of Israel, Haifa, Israel.

J. Brichard: Solvay et Cie., Neder over Hembeck, Bruxelles, Belgium.

G. G. Cameron: University of Aberdeen, Old Aberdeen, Scotland.

K. Camman: Technische Hochschule, München, W. Germany.

R. Cini: Università di Firenze, Firenze, Italy.

H. Collins: Cahn Instruments Co., Ltd., Dartford, Kent, England.

J. Cook: Ministry of Technology E. R. D. E., Waltham Abbey, Essex, England.

P. A. Cutting: Gas Council, London, England.

A. W. Czanderna: Clarkson College of Technology, Potsdam, New York, U. S. A.

R. J. Davidson: Anglo-American Research Laboratories, Johannesburg, South Africa.

M. A. Day: Imperial Chemical Industries, Ltd., Billingham, Teesside, England.

W. Dengler: Institut für Chemische Verfahrenstechnik, Stuttgart, West Germany.

D. Dollimore: University of Salford, Salford, Lancs., England.

A. M. J. Duymelinck: Eindhoven University of Technology, Eindhoven, Netherlands.

H. D. Duinker: N. V. Philips, Eindhoven, Netherlands.

D. Durand: Laboratoire de Minéralogie, Lyon, France.

K. M. Dijkema: Eindhoven University of Technology, Eindhoven, Netherlands.

A. Engberg: Danmarks Tekniske Højskole, Lyngby, Denmark.

J. Erkelens: Unilever Research Lab., Vlaardingen, Netherlands.

H. L. Eschbach: Euratom, Geel, Belgium.

P. Gaskins: Cahn Division, Ventron Instruments Corp., Paramount, Calif., U. S. A.

Th. Gast: Technische Universität, Berlin, W. Germany.

P. Glaude: Euratom, Petten, Netherlands.

S. J. Gregg: Brunel University, London, England.

F. Grønlund: University of Copenhagen, Copenhagen, Denmark.

E. A. Gulbransen: Westinghouse Research Lab., Pittsburgh, Pa., U. S. A.

D. Hacman: Balzers A. G., Liechtenstein.

D. Hillecke: Technische Universität, Clausthal-Zellerfeld, W. Germany.

F. B. Hugh-Jones: The Scientific Instrument Centre, London, England.

K. Keller: Mettler, Greifensee, Zürich, Switzerland.

M. G. Kennerly: Unilever Research Laboratory, Isleworth, Mddx., England.

G. P. Kerr: University of Aberdeen, Old Aberdeen, Scotland.

E. Knothe: Sartorius-Werke GmbH, Göttingen, W. Germany.

A. Kockel: Ruhr-Universität, Bochum, W. Germany.

H. Kreft: Sartorius-Werke GmbH, Göttingen, W. Germany.

W. Krückels: Institut für Chemische Verfahrenstechnik, Stuttgart, W. Germany.

H. van Leeuwen: Eindhoven University of Technology, Eindhoven, Netherlands.

J. A. van Lier: Union Carbide Corp., Parma, Ohio, U. S. A.

J. B. Lightstone: Union Carbide Corp., Tonawanda, N. Y., U.S. A.

B. C. Mansfield: English Electric Co., Ltd., Stafford, England.

C. H. Massen: Eindhoven University of Technology, Eindhoven, Netherlands.

B. McEnaney: Bath University, Bath, England.

J. P. de Mey: Eindhoven University of Technology, Eindhoven, Netherlands.

P. Michel: Laboratoire de Minéralogie, Lyon, France.

P. J. Møller: University of Copenhagen, Copenhagen, Denmark.

H. Moret: Euratom, Geel, Belgium.

W. H. Noakes: Cahn Instruments Co., Ltd., Dartford, Kent, England.

A. Pebler: Westinghouse Research Laboratories, Pittsburgh, Pa., U. S. A.

J. A. Poulis: Eindhoven University of Technology, Eindhoven, Netherlands.

J. Reinecke: Sartorius-Werke GmbH, Göttingen, W. Germany.

P. Reijnen: N. V. Philips, Eindhoven, Netherlands.

E. Robens: Battelle Institut e.V., Frankfurt/Main, W. Germany.

P. Sanmartin: Montecatini-Edison S. P. A., Porto Marghere, Italy.

P. Schmider: Technische Universität, Clausthal-Zellerfeld, W. Germany.

P. Schneider: Selb/Bayern, W. Germany.

H. Seiffert: Institut für Radiochemie, Karlsruhe, W. Germany.

J. C. Stouthart: Eindhoven University of Technology, Eindhoven, Netherlands.

S. J. D. van Stralen: Eindhoven University of Technology, Eindhoven, Netherlands.

B. Sülzer: Bergbau-Forschung, Essen-Kray, W. Germany.

E. Swaan: Eindhoven University of Technology, Eindhoven, Netherlands.

F. J. van Tongeren: Eindhoven University of Technology, Eindhoven, Netherlands.

M. Torrini: Università di Firenze, Firenze, Italy.

J. G. Vandenbroeck: Ers. H. V. L., Bruxelles, Belgium.

M. Verduin: Eindhoven University of Technology, Eindhoven, Netherlands.

R. Verouden: P. M. Tamson N. V., Zoetermeer, Netherlands.

M. W. M. Wanninkhof: N. V. Philips, Eindhoven, Netherlands.

L. Weickhardt: Sartorius-Werke GmbH, Göttingen, W. Germany.

H. G. Wiedemann: Mettler Forschungs Laboratorium, Greifensee, Zürich, Switzerland.

P. J. Wolff: Technological University Twente, Enschede, Netherlands.

P. Zuidema: N. V. Philips, Eindhoven, Netherlands.

Use of a Microbalance for the Determination of the Mass of Oxygen Reacting during the Oxidation of Thin Films of Binary Alloys

M. Boudeulle, D. Durand, and P. Michel

Laboratoire de Mineralogie
Lyon, France

ABSTRACT

The evolution of thin metallic layers during the oxidation processes is studied by means of electron diffraction. Comparison with known compounds, previously obtained in the bulk state, enables determination of the formula and hence the structure of the new compounds. The authors have tried to find the chemical formula of such products — in this case oxides of binary alloys — in a more scientific way. For this purpose they used a microbalance to determine directly, first the mass of the thin metallic layers and then the mass of oxygen reacting on them.

APPARATUS

The Cahn RG electrobalance used is placed in its pyrex bottle. This is fixed to a massive wall in a basement room (Fig. 1). The geometry of the tubes and of the sample support as well as the pressure conditions had been selected according to the results of Cahn and Schultz,[1,2] and Katz and Gulbransen.[3] The sample and the counterweight are suspended in quartz tubes (50 cm long, 5 cm in diameter). Airtightness is obtained by using O-ring seals.

The temperature is determined by a Pt/Pt-Rh thermocouple, fixed on the inner wall of the sample tube at the sample level (see Fig. 1). This tube is heated by a classical furnace, closed at both ends, the reaction temperature always being limited to 600 C. The suspension wires and stirrups are made of platinum.

By adjusting carefully the gas admission valve and the rotating pump flow, we keep the pressure constant at 100 torr, during all the operations of adjustment; in this way the recorder noise did not exceed 2 μg.

SAMPLES

The substrates are thin glass plates (area 400 mm^2, thickness 0.2 mm, mass <180 mg), previously vacuum annealed for two hours

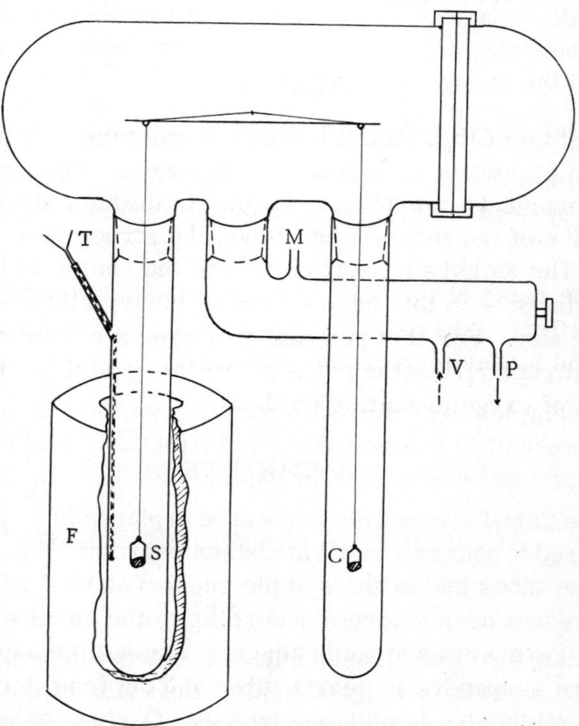

Fig. 1. Balance setting. T) Thermocouple; M) manometer; V) gas admission valve; P) pump; S) sample; C) counterweight; F) furnace.

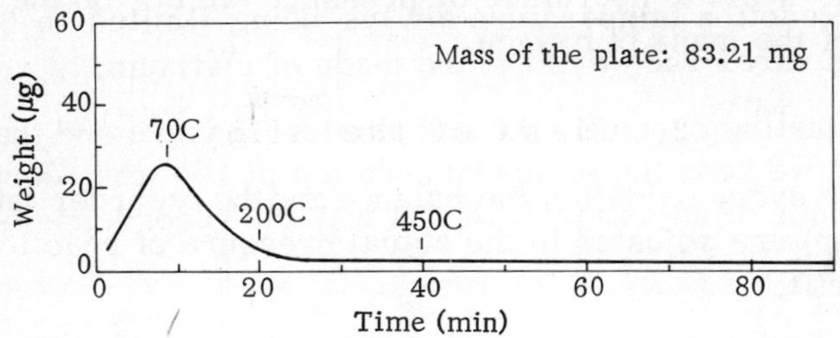

Fig. 2. Heating of a glass substrate at a pressure of 100 torr.

at 200 C, to prevent excess degassing and to give the sufficient stiffness during heating. They are pickled with sulfuric acid and washed with benzene. The behavior of the annealed plates is studied with the microbalance (Fig. 2). Three different parts may be noticed on the recorder curve:

First, the weight increases (about 30 μg) up to 70-80 C.

Then a slow decrease follows (about 25 μg). This loss may be caused by degassing of the quartz tube and resulting TMF effects.*

Then the mass of the plate remains constant during heating. These effects take place long before the beginning of the oxidations and so never interfere with them.

The samples are films of binary alloys, 1000 Å or more thick, vaporized in a good vacuum. Films are checked by x-ray fluorescence and electron microscopy (diffraction).

ACCURACY OF THE MEASUREMENTS

1. Procedure

The annealed and cleaned substrates are weighed. After vaporization, the sample is weighed again: the mass of the alloy is obtained by subtraction. We recorded directly the sample mass variations during the oxidation reactions and the cooling process

* The authors are greatly indebted to Dr. A. W. Czanderna for this explanation.

in order to avoid temperature or pressure effects on the determination of the mass of oxygen.

2. Mass of the Alloy

Before every oxidation the balance and the recorder (after long warming up) are adjusted to the actual pressure of reaction, to detect any drift.

Possible errors in the mass determination of the alloy are of two kinds:

A. Setting up and taking down of the stirrup. This experiment was done many times with a glass plate. The mechanical and electrical stability of the balance makes the mass difference between two settings never exceed 8 μg.

B. Contamination layers. Studies by electron diffraction of alloy films left for months in the open air never show the presence of oxides. One single layer of oxygen ions, on an area equal to that of the sample layer, has a mass of about 0.27 μg. If the contamination film is 50 Å thick, it means, because of the oxide structure, that only 10 such layers are involved in all. So the mass of the contaminating oxygen is certainly under 3 μg.

The error in the mass of the alloy, that mass always being more than 300 μg, never exceeds 12 μg, or 4%.

3. Mass of Oxygen

This mass is obtained directly from the recording — the noise is weak — between two points at the same temperature. Taking into account the contamination layer previously fixed, the error on the mass of oxygen is about 4 μg.

Two series (I and II) of experiments have been carried out.

I. Preparation of the Ferrite $NiFe_2O_4$ from $NiFe_2$ Alloy. These oxide films were required for further investigations on the Faraday effect. The film begins to oxidize at 300 C. At 450 C, the oxidation process is completed after 3 hours. No parasitic phenomenon is observed for the film itself (Fig. 3).

The calculation of the mass of absorbed oxygen shows that the oxidation of the film is complete and has, indeed, given the ex-

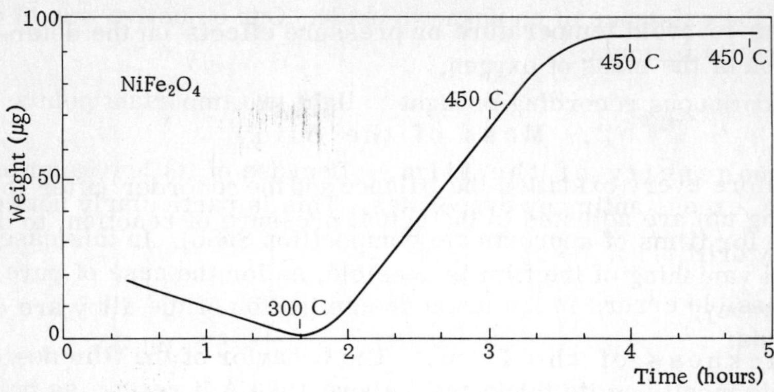

Fig. 3. Oxidation of $NiFe_2$ alloy at a pressure of 100 torr.

pected ferrite in accordance with the optical and electron microscopic observations (Table I).

Additional measurements of surface area and thickness (Tolansky) made possible the calculation of the density of the thin layer. Thus the electrobalance is a valuable implement in film-density studies.[4]

<u>II. Determination of the Formula of an Unknown Oxide of the Alloy SnSb.</u> Mixtures of antimony and tin are catalysts used in the making of acrylonitryl. Oxidation of SnSb in the open air at 600 C has

Table I

Parameter	Thin film I	Thin film II
Thickness, Å	1050 ± 20	1425 ± 20
Surface area, cm^2	3.98 ± 0.02	4.20 ± 0.02
Mass of alloy, μg	200 ± 8	283 ± 10
Film density, g/cm^3	4.73	4.78
Mass of oxygen, μg:		
measured	75 ± 3	105 ± 4
calculated	75	107

shown the existence of an unknown oxide. Our objective was to determine its formula.

Continuous recording brought to light two important points:

Homogeneity of the film. Because of its low vapor pressure, excess antimony evaporates. This is particularly noticeable for films of approximate composition $SnSb_2$. In this case total vanishing of the film is possible, as for the case of pure antimony.[5]

Thickness of the film. The behavior of the film depends mostly on its thickness. Above 1000 Å it reacts, as bulk products do, with greater stability. But the oxidation rate is never related to this factor.

SnSb alloy oxidation begins at 350 C. It proceeds gradually (1.2-1.4 μg/min) and is completed after 1 hour at 590° (Fig. 4). Exact determination of the mass of fixed oxygen has enabled us to determine the chemical formula of the unknown oxide of which the structure by means of electron diffraction was obtained. The formula $SnSbO_4$ appeared to fit the results of the measurements.

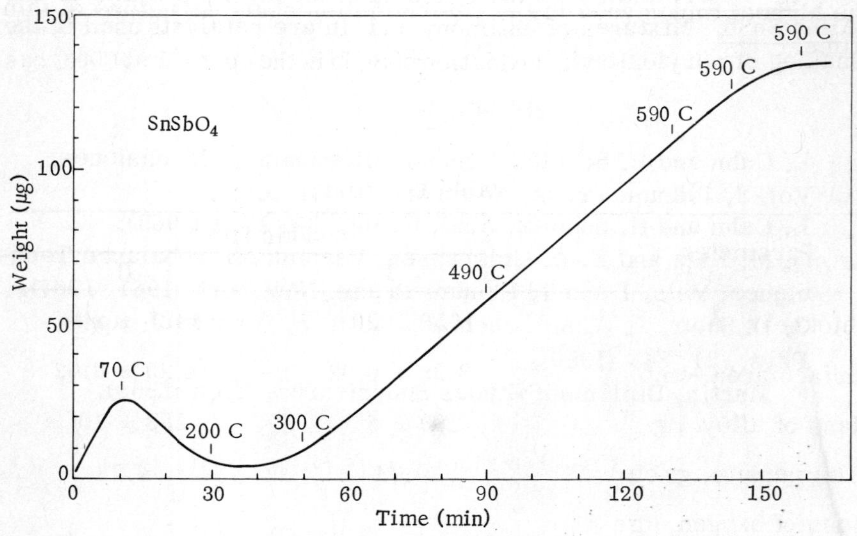

Fig. 4. Oxidation of SnSb alloy at a pressure of 100 torr.

Table II

Oxide formula	Mass of alloy, μg	Mass of oxygen, μg	
		measured	calculated
$SnSbO_4$	310 ± 10	84 ± 5	82
	470 ± 12	124 ± 4	125
	490 ± 12	128 ± 4	130
$SnSb_2O_6$	690 ± 12	120 ± 5	121

$SnSb_2$ alloy films of more then 1000 Å in thickness give an oxide, the formula of which is $SnSb_2O_6$. Because of their thickness it is impossible to check these films by electron diffraction (Table II).

CONCLUSION

The microbalance offers a valuable contribution to the study of compounds obtained in the form of thin films, for it makes possible a precise determination of their stoichiometric formula. For the crystallographer using electron diffraction this is a matter of the highest importance in the determination of the structure of thin films.

REFERENCES

1. L. Cahn and H. Schultz, Vacuum Microbalance Techniques, Vol. 3, Plenum Press, New York (1963), p. 29.
2. L. Cahn and H. Schultz, Anal. Chem., 35, 1729 (1963).
3. O. M. Katz and E. A. Gulbransen, Vacuum Microbalance Techniques, Vols. 1 and 4, Plenum Press, New York (1961, 1964).
4. G. D. Scott, T. A. McLauchlan, and R. S. Sennett, J. Appl. Phys., 21, 843 (1950).
5. F. Martin, Diplôme d'Études Supérieures, Lyon (1968).

Static Magnetic Susceptibility Measurements on Small Solid Samples

A. Van den Bosch

Static Magnetic Susceptibility Measurements on Small Solid Samples

A. Van den Bosch

Solid State Physics Department
S. C. K. - C. E. N. Mol, Belgium

ABSTRACT

An apparatus for the precise measurement of the magnetic susceptibility of small solid samples has been described in Volume 5 of the Vacuum Microbalance Techniques series. The striking part of the setup is a vacuum microbalance with a sensitivity of about 10^{-4} dyne. The aim of this contribution is to emphasize the need for such a balance. This is achieved by discussing measurements. In particular an analysis of the apparent susceptibilities of an impure MgO crystal is given, leading to a value of the diamagnetic susceptibility more negative than the one found in the literature. Furthermore, some data are reported on the paramagnetic compound Cs_2MnCl_4.

I. INTRODUCTION

The principle of the measurement of the static magnetic susceptibility is more than a hundred years old. Present-day research in this field therefore is characterized by the modern possibilities with respect to the construction of an apparatus and to the preparation of new materials. The author previously described[1] an apparatus designed to measure the susceptibility of small solid samples. The main part of this setup is an automatized torsion balance which is used to measure the force exerted on a specimen by a nonuniform magnetic field H. The coordinate system has been chosen so that the z-component of the force is measured. With ρ

the density of the sample and V its volume, the z component of the force acting on the sample is given by

$$F_z = \rho \int_V \left[[\chi_1 H_x \left(\frac{\partial H_x}{\partial z}\right) + \chi_2 H_y \left(\frac{\partial H_y}{\partial z}\right) + \chi_3 H_z \left(\frac{\partial H_z}{\partial z}\right)\right] dv \tag{1}$$

where χ_1, χ_2, and χ_3 are the principal values of the susceptibility. Equation (1) shows that a high-sensitivity force detector will make it possible to measure small samples with low susceptibility. The aim of this contribution is to emphasize, while discussing some of our results, the need for a vacuum microbalance having a sensitivity of about 10^{-4} dyne.

II. THE SHAPE OF THE SPECIMEN

For an isotropic and sufficiently small sample, Eq. (1) reduces to

$$F_z = m\chi_w \varphi_w \tag{2}$$

where

$$\chi_w = \chi_1 = \chi_2 = \chi_3 \tag{3}$$

$$\varphi_w = H_x \left(\frac{\partial H_x}{\partial z}\right) + H_y \left(\frac{\partial H_y}{\partial z}\right) + H_z \left(\frac{\partial H_z}{\partial z}\right) \tag{4}$$

and

$$m = \rho V \tag{5}$$

An experimental investigation can show how good the approximation is. Force measurements were carried out for the apparatus considered in the vicinity of the point where, on the z axis, the absolute value of $H_x \cdot \partial H_x/\partial z$ exhibits a maximum.[1] This place w is designed by the set of coordinates ($x = 0$, $y = 0$, $z = z_w$). A shift of the probe sample from z_w to $z_w \pm 0.1$ cm causes the value of φ to decrease by 0.3%. A deviation in the y direction out of $y = 0$ by ± 0.3 cm results in the same relative reduction of φ. A displacement of ± 0.1 cm from $x = 0$ in the x direction causes the force to increase by 0.3%. This investigation, carried out with the aid of a small, nearly cubic, specimen of about 5×10^{-4} cm^3, results in the following statement. An accurately positioned sample, the dimensions of which do not exceed 0.2, 0.6, and 0.2 cm in the x, y, and z directions, respectively, exerts on the balance a force which, to within 0.3%, is independent of the shape of the specimen. This behavior is advantageous for some experimentalists because at the moment single crystals of some refractory materials are available only as chips.

In order to give some idea of the magnitude of the values involved in the measurements, the data on magnesium oxide are given here. MgO is an isotropic diamagnetic material, the susceptibility of which has been reported to be 0.25×10^{-6} emu/g.[2] At w, where $H_y = H_z = 0$, the value of H_x lies between 7 and 17 kOe. At the same place the gradient of the field in the z direction is, roughly speaking, one-tenth of H_x per cm. The values of φ_w consequently are situated between 3.4×10^7 and 0.5×10^7 Oe^2/cm. In the situation just described, a MgO sample of 3×10^{-2} g is expected to exert a force on the sample between 0.3 and 0.04 dyne. If a precision better than 1% is wanted, a force detector having a sensitivity better than 4×10^{-4} dyne is needed.

III. DIAMAGNETIC MATERIALS

For small samples the force-equation (2) holds. Here it is expressed as

$$\left[F_z / m = \chi_w \varphi_w\right]_{I_M} \tag{2'}$$

In this form the equation emphasizes the fact that for different specimens of one material for which χ_w is a constant the force per unit of mass also is a constant if φ_w is the same for each specimen. More easy is the measurement of I_M, the current energizing the electromagnet. I_M is practically unequivocally related to φ_w. This statement is deduced from the experiment wherein Eq. (2') has been shown to hold for pure lithium fluoride crystals. The reproducibility of the measurement of F_z/m at room temperature is represented in Table I by the coefficient of variation ξ:

$$\left[\xi = 100 \cdot \sigma_{(F_z/m)} / (F_z/m)\right]_{I_M} \tag{6}$$

In this expression the standard deviation is

$$\left[\sigma_{(F_z/m)} = \sqrt{\sum_{j=1}^{n} d_j^2 / (n-1)}\right]_{I_M} \tag{7}$$

Table I

I_M, A	1.8	1.4	1.0	0.8	0.6
F_z/m, dynes/g	13.51	10.72	6.24	4.10	2.36
ζ, %	0.3	0.4	0.6	1.2	0.6

where n is the number of specimens considered and where, for the j-th specimen, d_j is the deviation from the mean, given by

$$\left[d_j = (F_z/m)_j - \sum_{j=1}^{n} (F_z/m)_j/n\right]_{I_M} \tag{8}$$

Some refractory materials are not available without ferromagnetic impurities. If it is desirable to know the diamagnetism of the material itself, the apparent susceptibility is to be corrected for a contribution that is due to the impurity. Such a correction is carried out for a MgO sample following the method of Honda and Owen.[4] The latter is based on the fact that ferromagnetic materials, at a temperature below their Curie temperature, saturate in high magnetic fields. Further, it is assumed that the impurities behave as if they have macroscopic dimensions, and therefore should not be too much dispersed in the host lattice.

These assumptions can be accounted for by putting a constant for the magnetic moment of the impurity:

$$\left[m_i \chi_{wi} H_x = K = \text{constant}\right]_{H_x > 7\text{ kOe}} \tag{9}$$

The subscript i indicates that the symbols are related to the impurity. The apparent susceptibility of a sample is obtained by dividing the total force, $\Sigma F_z = F_z + F_{zi}$, by $m\varphi_w$. This procedure follows from Eq. (2). It results in

$$\chi_a = \Sigma F_z/m\varphi_w = \chi_m + K/mH_x \tag{10}$$

Equation (10) describes a straight line in the χ_a versus $1/H_x$ diagram. In the case where the magnetic susceptibility χ_m of the material is independent of the field one can obtain its value by proceeding as follows. The χ_a values are plotted versus $1/H_x$, then a straight line is drawn through the points. Extrapolated, the line gives χ_m at $1/H_x = 0$. This procedure has been applied for the MgO sample, which at the time it was measured was one of the purest available. The apparent susceptibility as obtained at three different temperatures is given in Fig. 1. The application of the corrections for the contributions of the ferromagnetic impurities results in a temperature-dependent susceptibility χ_m. This is not what is expected from the theory of the diamagnetism of ions, but it can be explained by the presence of paramagnetic impurities.

Dispersed transition-metal ions exhibit paramagnetism. If it is assumed that this obeys the Curie law, then the susceptibility of the material, which can be considered as the sum of a diamagnetic and a paramagnetic term, can be written as

$$\chi_m = \chi_d + \chi_p = \chi_d + P/T \tag{11}$$

P being the Curie constant. Equation (11) again represents a straight line, this time in a χ_m versus $1/T$ diagram. The same procedure as above now leads to the susceptibility of the pure material. The analysis given in Fig. 2 shows that $\chi_d = -0.38 \times 10^{-6}$

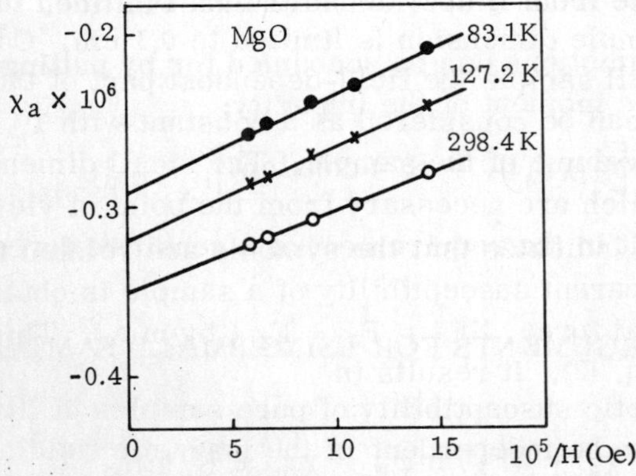

Fig. 1. The apparent susceptibilities of a MgO crystal (≃ 30 mg) are plotted versus the reciprocal of the magnetic field. The measurements were carried out at three different temperatures.

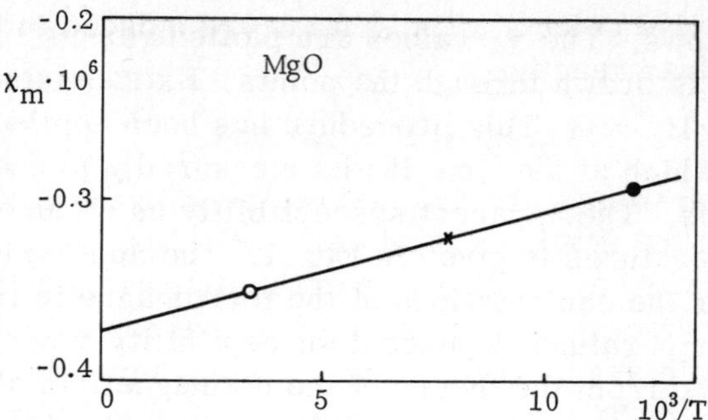

Fig. 2. The apparent susceptibilities of a MgO crystal (≃ 30 mg), corrected for a contribution which is due to ferromagnetic impurities, plotted versus the reciprocal of the absolute temperature.

emu/g. This value is much more negative than the one reported in the literature.[2]

In the case of MgO, the apparent susceptibility clearly depends on the field strength and therefore also on the coordinates of its volume elements dv. The field-dependent part of χ_a is inversely proportional to the field. The field in the sample however does not deviate from H at w by more than 1%, when in the z direction the sample dimension is limited to 0.1 cm. Consequently, for such a small sample the field-dependent part of the apparent susceptibility can be considered as a constant with 1% uncertainty over the total volume of the sample. The small dimensions of the sample, which are necessary from the point of view of the analysis, result in the requirement of a sensitive force detector.

IV. ARGUMENTS FOR USING SMALL SAMPLES

The magnetic susceptibility of pure samples of lithium fluoride was found to be independent of the magnetic field and of the temperature (between 60 and 300 K) within 0.5%. Neutron-irradiated LiF, however, exhibits a temperature-dependent susceptibility.[5] This can be explained as composed of two parts: 1) the diamagnetic part of the normal LiF matrix, and, 2) the temperature-dependent part which is attributed to isolated unpaired electrons bound to structural defects. The defects are induced in the crystals by the nuclear reaction

$$^{6}\text{Li}(n, \alpha)^{3}\text{H} + 4.8\ \text{MeV}$$

The mean penetration of the thermal neutrons in the natural material, in which the isotopical abundance of ^{6}Li is 7.4%, is about 0.24 cm. Crystals having large dimensions as compared to this length will be inhomogeneously damaged. One can calculate that, in order to obtain samples that are irradiated homogeneously to better than 1%, the thickness of a plate of LiF should be less than 0.05 cm. Here again a thin sample is needed, and consequently a highly sensitive force detector is recommended.

Neutron irradiation may furthermore give rise to radioactive specimens. Even in the case where the pure material does not activate, the presence of small concentrations of impurities may result in a troublesome activity after the sample has been irradiated by a high neutron dose. Instead of constructing an elaborate system for the remote mounting of a large and, in view of Eq. (1), shaped specimen on the balance, it is far easier to use small samples.

V. TRANSITION-METAL COMPLEXES

For high concentrations of transition-metal ions the paramagnetic susceptibility may become so large that even smaller samples than in the case of the diamagnetic substances can be used. Measurements were carried out on single crystals of K_2NiF_4 of about 1 mg.[6] The values of the susceptibilities, obtained at different temperatures for different orientations, are in good agreement with those reported in the literature. This result indicates that these small samples are still representative of the bulk. With the aid of the same apparatus small single crystals of K_2CoF_4 were investigated.[7] Their magnetic behavior was found to be analogous to that of the K_2NiF_4, except for a larger anisotropy in the temperature region above the Néel temperature. In the low-temperature region the agreement with data to be found in the literature is not so good. The discrepancy can be explained by assuming that our small samples are more perfect, which assumption is plausible, since it is difficult to grow large single crystals of these materials. Single crystals of K_2CuF_4 of about 1 mg were also investigated,[7] but no data were found in the literature for comparison.

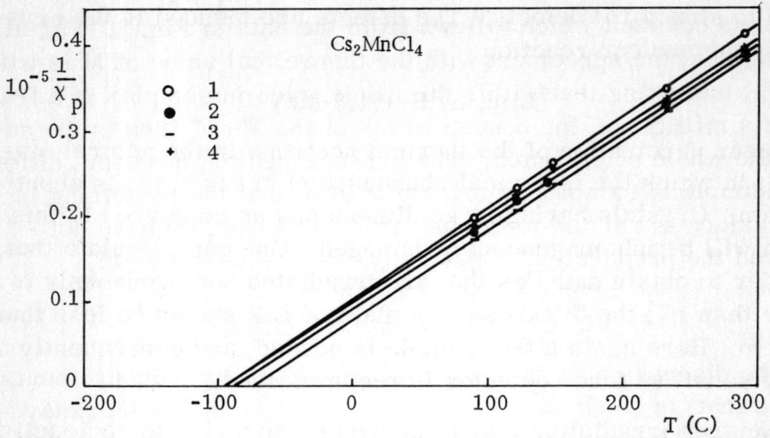

Fig. 3. The reciprocal of χ_p for Cs_2MnCl_4 samples of $\simeq 1$ mg, plotted against the absolute temperature.

Of Cs_2MnCl_4 only powder specimens were obtained. Measurements were also carried out on samples of less than 1 mg of this material. The value of the susceptibility obtained at room temperature is of the same order of magnitude as that given by Asmussen.[3] In Fig. 3 the inverse of the paramagnetic susceptibility is plotted versus the temperature for four different samples taken out of one batch of Cs_2MnCl_4. The difference of the susceptibility of two different samples is in some cases as high as 12%. This is not due to the inaccuracy of the measurements, so one may suspect the preparation of the sample. A badly mixed $MnCl_2$ + 2CsCl powder should result in a nonhomogeneous distribution of the Mn^{++} ions in the batch. However, in the $1/\chi_p$ versus T diagram a straight line can be drawn through the points related to the same sample. Consequently, in all specimens measured the paramagnetic susceptibility obeys the Curie — Weiss law. Since the magnetic moment for Mn^{++} is known, one can calculate the molecular weight M of a sample from the relation

$$M = \frac{N}{\chi_p} \cdot \frac{\mu_B^2 g^2 S(S+1)}{3k(T-\theta)} \tag{12}$$

where N is Avogadro's number, μ_B the Bohr magneton, g the Landé factor, S the total spin of the ion, k the Boltzmann constant, and θ

the Weiss constant which follows from the data in Fig. 3. For all the samples, the agreement with the theoretical value of M is within 2.5%, indicating that within this limit, even in samples of a fraction of a milligram, the concentration of the Mn^{++} ions is the same as in the whole of the batch. The susceptibilities, however, differ from sample to sample, so that we believe that the scattering in the χ values is related to the fact that, until now, we have not achieved the preparation of perfectly crystalline material.

VI. CONCLUDING REMARKS

The discussion on the magnetic susceptibility measurements showed that, in some cases, it is convenient to use the Faraday method, in which the force on a small sample in a nonuniform field is measured. The small dimensions of the MgO sample allowed for taking in Eq. (2) the field-factor φ_W independently of the shape of the specimen. The same equation shows that φ_W, which is limited for practical reasons, results in small forces on small samples with low values of the susceptibility. Consequently, a very sensitive force detector is needed. This detector, in the case of LiF, makes it possible to measure on thin crystals, which can be irradiated homogeneously with thermal neutrons. For neutron-irradiated specimens in general one can say that the less material is treated, the less troublesome the radioactivity is to deal with. Furthermore, the possibility to measure small samples enables one to check the homogeneity of more voluminous batches as shown by the work with Cs_2MnCl_4. The fact that the values obtained on K_2NiF_4 agree with those in the literature indicates that the present apparatus[1] is reliable. Out of a nonmagnetic Cu—Ni alloy[8] a pan of about 50 mg was made for the balance. The lowest susceptibility measured on this material with the aid of the setup was 4×10^{-9} emu/g. Consequently, forces equivalent to about 1 μg had to be registered. The possibility of measuring in vacuum allowed investigation of highly hygroscopic materials without major difficulties.

ACKNOWLEDGMENTS

The work has been performed under the auspices of the association S. C. K. - R. U. C. A. We also are grateful to Dr. E. Legrand, who kindly lent us the transition-metal-complex samples.

REFERENCES

1. A. van den Bosch, An apparatus for the precise measurement of magnetic susceptibilities of solids, in: Vacuum Microbalance Techniques, Vol. 5, K. H. Behrndt, ed., Plenum Press (1966), p. 77.
2. C. D. Hodgman, ed., Handbook of Chemistry and Physics, Chemical Rubber Publishing Co., Cleveland, Ohio (1955), p. 2393.
3. R. W. Asmussen, The magnetic criteria for bond type in complex compounds, Proc. Symp. Coordination Chem., Copenhagen (1953), p. 27.
4. J. J. Donoghue, NAA-SR-117 (North American Aviation, Inc., U. S. A.) (1953).
5. R. T. Bate, and C. V. Heer, J. Phys. Chem. Solids, 7, 14 (1958).
6. A. van den Bosch, Meded. BNV IV, 371 (1965).
7. A. van den Bosch and E. Legrand, Bull. Belg. Phys. Soc., V, 223 (1967).
8. E. W. Pugh, Rev. Sci. Instr., 29, 1118 (1958).

Measurements of Magnetic Thin-Film Parameters by the Use of an Automatic Microbalance

Judith Bransky, I. Bransky,* and A. A. Hirsch

Department of Physics
Technion — Israel Institute of Technology
Haifa, Israel

ABSTRACT

This paper describes the application of a Cahn RG electrobalance for measurements of the magnetic parameters of thin ferromagnetic films. By a suitable conversion of the microbalance into a torque-measuring device, the static hysteresis loops and torque curves can be directly recorded. From the recorded data the main magnetic properties of the films, such as saturation magnetization, remanence, coercive force, and the induced anisotropy constant can be evaluated with high accuracy. Examples of hysteresis loops of various ferromagnetic films with thicknesses ranging between 200 and 2000 Å are given.

I. INTRODUCTION

The evaluation of the magnetic parameters of thin ferromagnetic films by the conventional quasistatic "change-of-flux" methods presents a severe experimental problem due to the small quantity of material and its geometry. In 1963, Humphrey and Johnston[1] showed how the main magnetic parameters such as saturation magnetization, remanence, coercive force, and anisotropy constant of very thin films could be measured statically by means of a sensi-

* Now at Wright-Patterson AFB, Dayton, Ohio, USA.

tive torque meter. A year later Torok et al.[2] suggested a method for plotting the M versus H hysteresis loop of thin films with essentially the same instrument.

In the course of investigating the magnetic properties of thin ferromagnetic films it was found that such measurements could be performed conveniently by conversion of a sensitive commercial electrobalance to a torque meter. The aim of the present paper is to describe the modification and the technique used for such measurements.

When a ferromagnetic film is placed in a horizontal magnetic field H, with its normal parallel to the field, a torque proportional to the vector product of the film magnetization M_s and the field H is exerted on the film. Because of the strong demagnetization field it is usually assumed that the film magnetization of M_s lies exclusively in the film plane. The vertical component of the torque is therefore given by

$$L_z = Hm_s \cos\alpha \tag{1}$$

where m_s, the magnetic moment of the film, is equal to M_sV, V being the film volume; $m = m_s \cos\alpha$ is the horizontal component of the magnetic moment. A nickel film ($M_s = 484$), 100 Å thick and 2 cm^2 in surface area, will sense, in a field of 100 Oe, a maximum torque of about 10^{-1} dyne-cm. For such measurements a sensitivity of about 5×10^{-3} dyne-cm is required. The rated sensitivity of the Cahn RG electrobalance[3,4] is 0.2 μg, and the weighing loops are 12 cm apart. Thus, in optimal conditions, the balance is capable of measuring a torque of 10^{-3} dyne-cm.

II. THE CONVERSION OF THE BALANCE INTO A TORQUE MAGNETOMETER

The conversion of the balance into a torque meter has been described in detail.[5] The balance is turned with its front downward in order to measure the vertical torque. The torque is transmitted to the balance beam through a straight "soda straw" (60 cm long) and a light phosphor bronze frame, in turn supported on a small flat sapphire by means of a steel needle. The sapphire is mounted on a screw and may be elevated until the arms of the frame are in contact with those of the balance (Fig. 1). The whole system is enclosed in a glass vacuum chamber evacuated to 10^{-5} torr.

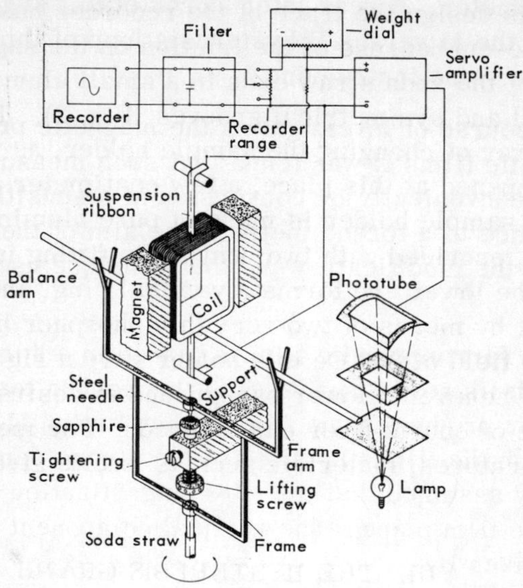

Fig. 1. The moving-coil system of the balance in the rotated position, the torque-transmitting frame with its support, and a block diagram of the electrical circuit.

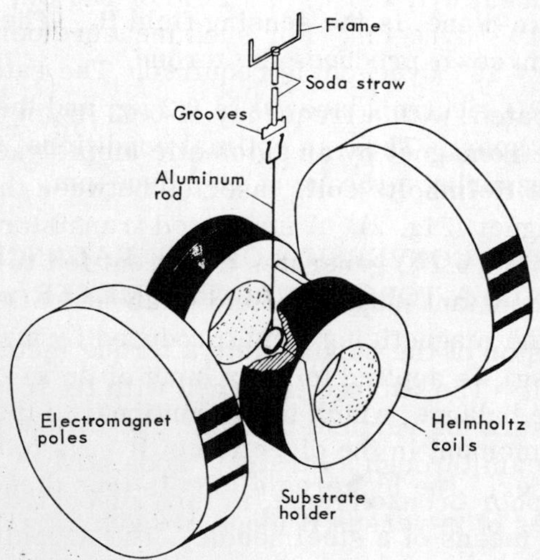

Fig. 2. The perpendicular magnetic fields: poles of the iron-core electromagnet and Helmholtz coils. The substrate holder and transmitting rod in detail.

With this design the trace of the recorder has a noise level of 3×10^{-3} dyne-cm because of the friction on the sapphire. The lower part of the soda straw ends in a small aluminum strip with two identical and symmetrical grooves (Fig. 2). This provides a convenient way of changing the sample holder, as the vacuum chamber can be opened at this place, a few centimeters above the pole edges. The sample holder is made of pure aluminum wire 1 mm in diameter, provided with two bent arms fitting into the two grooves. The lower end forms a vertical ring, and the substrate is fixed to it by means of two very thin phosphor bronze spring strips. In a field of 800 Oe this holder gave a signal of about 10^{-2} dyne-cm. Whenever a lower background is desired, a sample holder made of quartz fiber can be used.[1] For measurements at room temperatures, materials such as Lucite give good results.

III. THE HYSTERESIS GRAPH

In order to record directly the static hysteresis loop of a thin ferromagnetic film, two perpendicular magnetic fields are required.[2] The first field, parallel to the film plane, serves to switch the magnetization in the film plane. This magnetizing field H_m is a low-frequency-variation field. The second field, perpendicular to the film plane, is the sensing field H_s. The interaction between H_s and $m_s \cos \alpha$ produces the torque.

H_m is generated, with a frequency of 0.02 Hz, in an "Oerlikon C_1" rotating electromagnet by an automatic amplidyne. H_s is supplied by a pair of Helmholtz coils inserted between the flat poles of the electromagnet (Fig. 2). A stabilized transistor power supply, regulated to within $\pm 0.1\%$, generates a current fed to the coils, giving rise to a constant magnetic field of about 80 Oe. A voltage proportional to the magnetizing field, produced by a sensitive Halltron (0.34 V/A kg), is applied to the x input of an $x - y$ recorder. The output of the balance, which is proportional to the projection of the magnetic moment in the direction of H_m, is fed to the y input of the recorder. The hysteresis loop is thus directly recorded. Various examples of m versus H loops are given in Fig. 3.

In Fig. 3a the loop of an iron film deposited on a glass substrate by the conventional method is given. This film is 1500 Å thick and 2 cm^2 in surface area.

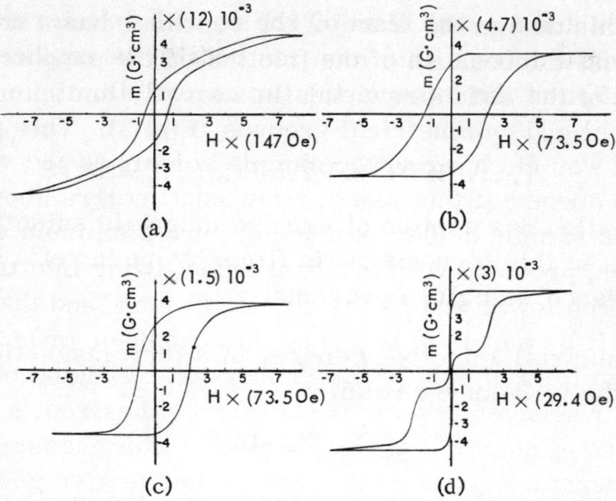

Fig. 3. Magnetization hysteresis loops of different materials: (a) an iron film 1500 Å thick deposited on a glass substrate measured at 300 K; (b) a cobalt film 700 Å thick deposited on a CoO single crystal measured at 270 K (c) a nickel film 500 Å thick deposited on a NiO single crystal measured at 300 K; (d) a composite film of Permalloy 370 Å thick, with an intermediate layer of SiO, 1000 Å thick, and a cobalt film 120 Å thick.

The two following loops exhibit unidirectional anisotropy due to exchange interactions. Figures 3b and 3c represent hysteresis loops of ferromagnetic films deposited on antiferromagnetic single crystal substrates of NiO and CoO having Néel temperatures of 520 K and 300 K, respectively.

The loop in Fig. 3b is the hysteresis of a cobalt film about 700 Å thick deposited on CoO with surface parallel to the $\{111\}$ plane. The loop was recorded while the sample was at 270 K. Figure 3c represents the hysteresis loop of a nickel film, 500 Å thick, deposited on NiO with surface parallel to the $\{100\}$ plane; the sample was at room temperature. Both loops are seen to be shifted along the x axis due to the ferro — antiferromagnetic interaction.[6] A detailed account of this work is being prepared for publication.

Figure 3d is an anomalous loop with two coercive forces of a composed film. It represents the hysteresis of a Permalloy film

370 Å thick and a cobalt film of 120 Å, with an intermediate layer of SiO, 1000 Å thick. This loop can roughly be explained as a superposition of the loops of the two materials.

IV. ANISOTROPY MEASUREMENTS

A detailed description of induced magnetic anisotropy measurements in thin ferromagnetic films by means of a Cahn RG electrobalance was given previously.[5]

The uniaxial anisotropy energy of a ferromagnetic film is given by the well-known relation

$$E = K_u \sin^2\theta \tag{2}$$

where K_u is the anisotropy constant, and θ the angle between the easy direction and the magnetic moment m_s, of the film. When a magnetic field H is applied parallel to the film plane with an angle φ to the easy axis, the magnetic torque $m_s H \sin(\varphi - \theta)$ will be equal to $K_u \sin 2\theta$, as $L = dE/d\theta = K_u \sin 2\theta$.

In order to evaluate K_u the film is placed horizontally between the flat poles of the electromagnet. The whole magnet is uniformly rotated around the sample by a synchronous motor and a gear. The balance output is given to the y axis of the recorder, and a voltage proportional to the angle of rotation of the electromagnet is fed to the x axis. The torque curves are recorded in various magnetic fields, high enough to make $\sin(\varphi - \theta) \ll 1$. The coefficients of $\sin 2\theta$ are then calculated from each curve by Fourier analysis and extrapolated to $1/H \rightarrow 0$ to give K_u.

In this case the measured torque is much smaller than that of the film in the vertical position (see Section III), where the high demagnetizing energy forces the magnetic moment of the film to be perpendicular to the sensing field.

Since the constant K_u of a nickel film is of the order of 10^3 ergs/cm^3, the maximum torque of a film 1000 Å thick and 2 cm^2 in surface area amounts to 2×10^{-2} dyne-cm. By suspending the torque-transmitting assembly directly from the balance beam, the noise level is reduced to 2×10^{-4} dyne-cm and the torque curves can thus be recorded with high resolution, provided the measurements are carried out in a vacuum.

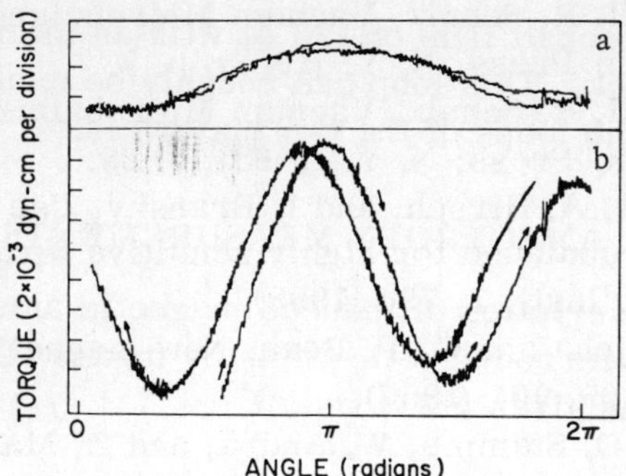

Fig. 4. Example of torque curves: (a) the "background," i.e., torque exerted by a field of 1000 Oe on a bare glass substrate; (b) a torque curve of a 1000-Å-thick nickel film in a field of 400 Oe, recorded in clockwise and counterclockwise directions.

It should be stressed that measurements of such high sensitivity can be severely affected by distortions caused by inhomogeneity of the magnetic field, misalignment of the film, and contributions of the sample holder giving rise to parasitic torque. The elimination of these effects is discussed in References 5 and 7.

In Fig. 4a the residual parasitic torque of a bare substrate is given. A typical sin 2θ curve for a nickel film 1000 Å thick is given in Fig. 4b.

ACKNOWLEDGMENT

The authors wish to thank Mr. I. Zviely for his devoted technical assistance.

REFERENCES

1. F. B. Humphrey and A. K. Johnston, Sensitive automatic torque balance for thin magnetic films, Rev. Sci. Inst. 34, 348 (1963).
2. E. J. Torok, D. C. Agouridis, A. L. Olson, and H. N. Oredson, Steady-state B — H hysteresis graph for thin ferromagnetic films using a torque magnetometer, Rev. Sci. Inst., 35, 1039 (1964).

3. L. Cahn and H. R. Schulz, Vacuum Microbalance Techniques, Vol. 2, Plenum Press, N. Y. (1962), p. 7.
4. L. Cahn and H. R. Schulz, Vacuum Microbalance Techniques, Vol. 3, Plenum Press, N. Y. (1963), p. 29.
5. J. Bransky, A. A. Hirsch, and I. Bransky, Use of commercial electric microbalance for highly sensitive torque measurements, J. Sci. Inst., 1, 790 (1968).
6. W. H. Meiklejohn and C. P. Bean, New magnetic anisotropy, Phys. Rev., 105, 904 (1957).
7. W. Schüppel, O. Stemme, W. Andrä, and Z. Málek, A method for the direct measurement of magnetic anisotropy in thin films, Fiz. Metal. Metalloved., 8, 837 (1959).

Activation Energies of the Decomposition of Poly(methyl α-phenylacrylate) from Static and Dynamic TGA

G. G. Cameron and G. P. Kerr

Department of Chemistry
University of Aberdeen
Old Aberdeen, Scotland

ABSTRACT

It is shown that good agreement between activation energies, E_a, as determined by the two methods occurs only when the mechanism of degradation is invariant with extent of decomposition. In the case of poly(methyl α-phenylacrylate), which decomposes by random initiation and unzipping to monomer, this condition only applies with low-molecular-weight material. In polymers which contain long chains a termination step reduces E_a. As the amount of termination diminishes with average molecular weight (i.e., with extent of reaction) E_a from dynamic TGA is an average. E_a from static TGA (initial conditions) does not suffer from this disadvantage.

I. INTRODUCTION

As the field of application of synthetic high polymers has widened there has been an increasing amount of interest in their thermal properties. For comparative purposes one of the most important parameters to be studied is the activation energy of decomposition, which in general varies directly with thermal stability.

The bulk of the literature data on the thermal decomposition of polymers derives from initial rates of volatilization measured under isothermal conditions. In order to obtain an activation energy for decomposition in this way, however, a series of separate measurements has to be made. This can often be time-consuming, as well as wasteful in material, and there has been a tendency recently to use dynamic thermogravimetric analysis (TGA) with linear heating rates, in attempts to obtain a set of kinetic data from a single TGA trace.

While dynamic TGA is a well-established technique in the study of the decomposition of inorganic salts, salt hydrates, and chelates and simple organic compounds, where the reactions are usually simple and unambiguous, the uncritical application of this technique to the study of polymer decomposition is unwise because the mechanism of decomposition may no longer be simple or invariant with temperature and extent of reaction. A recent review by Flynn and Wall[1] of the available mathematical methods of analysis of dynamic TGA data emphasizes, from a theoretical standpoint, the unreliability of values for the activation energy and order of reaction determined from a single TGA curve, even in some cases where a single Arrhenius expression is operative. Consequently the good agreement claimed to exist for several polymers between the activation energies from static (initial rates) and dynamic TGA may be fortuitous. Further doubt is cast on the validity of these comparisons since different experimental equipment and procedures, which in themselves may be sources of variation, have been used in the two types of experiment.

As part of a study of the pyrolysis of poly(methyl α-phenylacrylate) (PMPA), we have degraded identical polymer samples under both static and dynamic heating conditions in the same apparatus. By greatly reducing the possible effects of instrumental variables a more meaningful comparison of experimental activation energies from static and dynamic TGA can be drawn.

PMPA is particularly suitable for this study as the mechanism of decomposition is fairly simple, monomer being the only product of degradation.[2,3] The polymer samples, which were obtained from a number of sources, were all prepared by anionic synthesis, as the monomer does not homopolymerize by a free-radical route.[4] The samples were characterized by viscometry and osmometry.

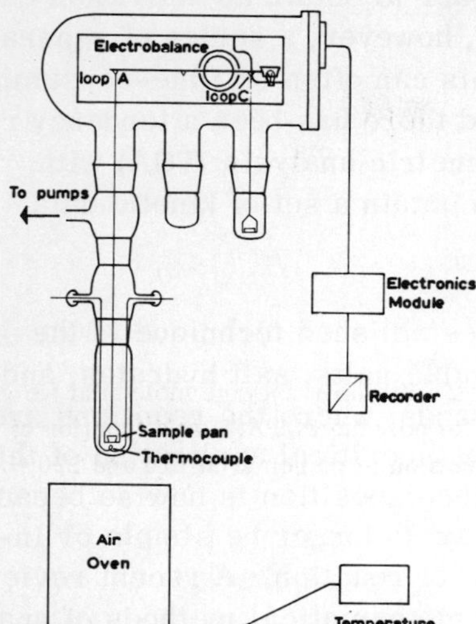

Fig. 1. Electrobalance installation.

II. ELECTROBALANCE INSTALLATION

The balance employed was a Cahn RG Electrobalance, which was mounted in the glass vacuum bottle supplied by the manufacturer (Fig. 1). The sample pan was a platinum crucible 1 cm in diameter. During degradations the sample pan was placed on a small copper stirrup which was suspended from loop A on the balance beam by a rigid glass fiber 60 cm long. The total weight of pan + stirrup + fiber was counterweighted on loop C, and the balance was calibrated to read from 0 to 10 mg on a Leeds and Northrup 1 mV recorder. Polymer samples weighed 9-10 mg.

The system was evacuated to better than 10^{-4} torr by a mercury-diffusion pump, backed by a rotary oil pump. The furnace was a fan-circulated air oven whose temperature was controlled by an F & M type 240 temperature controller and programmer, actuated by a thermocouple inside the oven.

The permanent copper — constantan thermocouple beneath the balance pan was calibrated to give the sample temperature during a run by carrying out a series of blank experiments with a second thermocouple embedded in a spot of silicone grease on the pan surface.

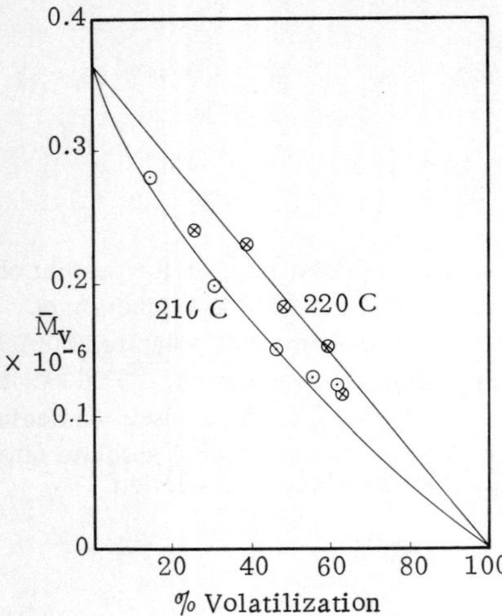

Fig. 2. Viscosity average molecular weight $\overline{M}_v$ for polymer PMPA(5) as a function of conversion to monomer at 210 and 220 C.

III. RESULTS AND DISCUSSION

A preliminary study[3] of the molecular weight changes in sample PMPA(5) during degradation showed that the degree of polymerization (DP) fell steadily during the reaction (Fig. 2) and that decomposition involved monomer formation following initiation by random homolytic scission of internal C — C backbone bonds. This behavior is analogous to that exhibited by anionically synthesized poly(methyl methacrylate)[4,5] and poly(α-methylstyrene),[6] and would be expected from the absence of active tertiary hydrogen atoms and labile chain ends in the polymer. These experiments also indicated a kinetic chain length (KCL) of ~1000 monomer units at 210 C.

All five polymer samples were degraded isothermally at four temperatures between 230 and 280 C in the electrobalance. The initial rates of volatilization were obtained by extrapolation of the rate curves to zero conversion, and Arrhenius plots of log (initial rate) versus reciprocal absolute temperature were drawn as shown in Fig. 3.

Temperature-programmed degradations were effected at a furnace heating rate of 3 deg min^{-1}; the actual rate of temperature increase in the sample pan was 2.77 deg min^{-1}. The starting temperature in all experiments was 100 C. The difference-differential me-

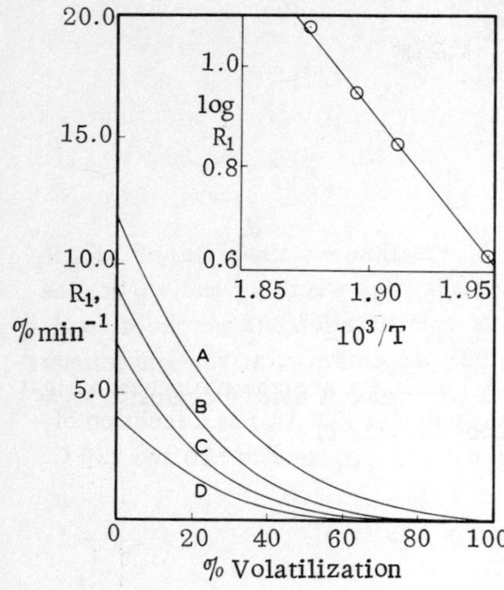

Fig. 3. Rate of volatilization R_1 (percent of original sample per min) as a function of conversion to monomer for polymer PMPA(1) during isothermal degradation. A) 260 C; B) 255 C; C) 250 C; D) 240 C. Inset: Arrhenius plot of log R_1 versus reciprocal absolute temperature.

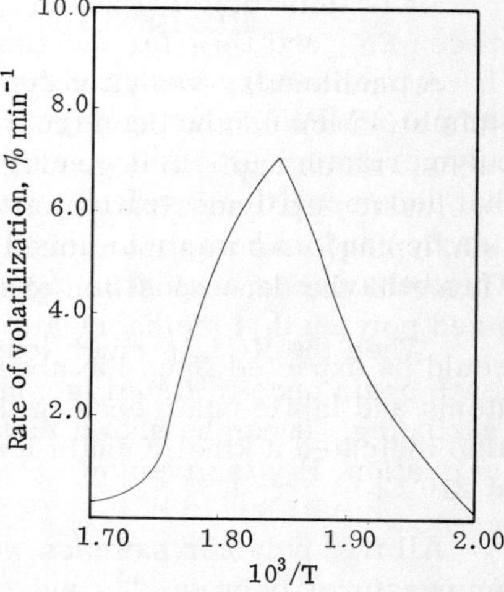

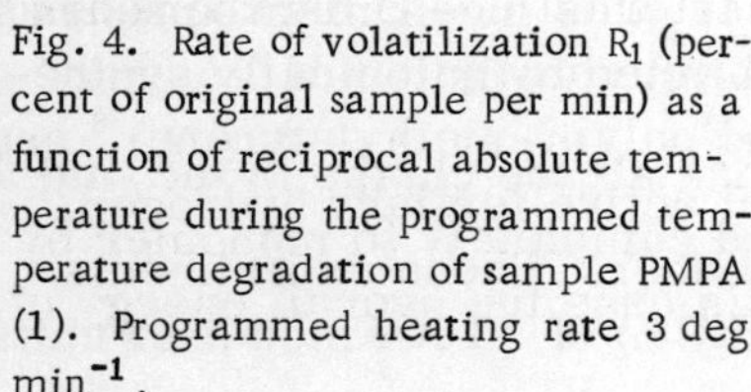
Fig. 4. Rate of volatilization R_1 (percent of original sample per min) as a function of reciprocal absolute temperature during the programmed temperature degradation of sample PMPA. (1). Programmed heating rate 3 deg min^{-1}.

thod of Freeman and coworkers[7,8] was used to determine activation energies (and orders of reaction) from the experimental volatilization versus time curves. Figures 4 and 5 show some typical rate data from TGA curves and Table I summarizes the activation energies from both static and dynamic measurements (E_{stat} and E_{dyn}).

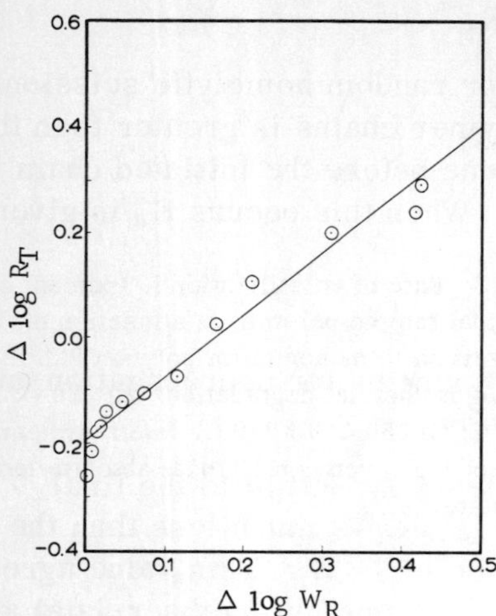

Fig. 5. Freeman — Carroll plot of $\Delta \log R_T$ versus $\Delta \log W_R$, where R_T and W_R are the rate of volatilization and weight of residual polymer, respectively, at absolute temperature T. $\Delta \log R_T$ is plotted against $\Delta \log W_R$ for constant $\Delta(1/T)$.

It is immediately obvious from Table I that the agreement between E_{dyn} and E_{stat} for the same polymer is poor, except for the lowest-molecular-weight polymer, PMPA(2). The reasons for this situation can be seen from the following energetic considerations,[9] which apply to degradation initiated at random, followed by exclusive depolymerization of the primary radicals to monomer (unzipping). As was mentioned earlier, this mechanism appears to describe the decomposition of PMPA very well.

When the KCL is much longer than all the chains in the sample, each chain once initiated is converted completely to monomer by unzipping. It can be shown that in this case the overall energy of activation, E_a, is given by

Table I. Molecular Weights and Activation Energies of Decomposition of Poly(methyl α-phenylacrylate) Samples

Sample	$\overline{M}_v$	$\overline{M}_n$	E_{dyn}	E_{stat}
PMPA(1)	474,000	30,000	43.3	28.3
PMPA(2)	17,000	17,000	79.3	80.6
PMPA(3)	50,000	15,000	51.3	39.7
PMPA(4)	47,000	12,000	68.4	60.8
PMPA(5)	366,000	140,000	66.1	56.2

$$E_a = E_1 \qquad (1)$$

where E_1 is the activation energy for random homolytic scission. When the average DP of all the polymer chains is greater than the KCL a termination step will intervene before the initiated chain is converted completely to monomer. When this occurs E_a is given by

$$E_a = \tfrac{1}{2}E_1 + E_2 - \tfrac{1}{2}E_3 \qquad (2)$$

where E_2 and E_3 are the activation energies for depropagation and bimolecular termination, respectively.

For PMPA(2), Eq. (1) applies from the initial to the final stages of degradation since the DP is always much less than the KCL. Thus $E_{stat} = E_{dyn} = 80$ kcal mole^{-1} $= E_1$. This value agrees well with literature values of E_1 for poly(methyl methacrylate) and poly(α-methylstyrene).[10] In this sample the mechanism is therefore invariant with extent of reaction, a single Arrhenius expression operates throughout, and dynamic TGA gives a meaningful activation energy.

For the other four samples, the agreement between E_{stat} and E_{dyn} is poor. These discrepancies arise because the higher molecular weights and wider molecular-weight distributions of these polymers introduce a measure of biradical termination into the degradation mechanism. The energetics of the process must therefore be intermediate between those represented by Eqs. (1) and (2), the contribution from each being determined by the proportion of chains capable of complete unzipping. Accordingly the net effect of the incorporation of a termination step into the degradation mechanism is to decrease E_a.

Furthermore, termination — which probably occurs by disproportionation at these elevated temperatures — can lead to the formation of unsaturated chain-ends.[11] Subsequent reinitiation of degradation at these terminal units should proceed with a lower activation energy than randomly initiated decomposition. This would further lower E_a. Thus the decremental effect of termination on E_a may be twofold.

As the molecular weight of the sample decreases with increasing conversion (Fig. 2), the effect of termination on E_a will steadily diminish also, and in the limit may vanish completely. Values of E_{dyn}, as determined by a single-curve method such as that of Freeman and collaborators, will therefore be procedural averages. Values of E_{stat} based on initial conditions will not reflect such changes, however, since they are based on instantaneous rate measurements at zero conversion. The above explanation also predicts that, for a degradation mechanism of the type occurring in PMPA, E_{dyn} will be greater than E_{stat}, except where the DP is $\ll$KCL. The results in Table I are in line with this prediction.

It is clear that the results from static measurements give a more realistic assessment of sample stability, although from a kinetic viewpoint only the result for PMPA(2) in Table I has any real significance, due to the unknown variations in the amount of termination occurring in the other samples. It should also be noted that for a sample in which DP is initially $\gg$KCL and to which Eq. (2) can be applied, E_{stat} from initial rates would be significant. On the other hand, if the DP diminishes sufficiently during degradation, so that a proportion of the chains are able to unzip completely at higher conversions, E_{dyn} averaged over the whole reaction range would be rather meaningless.

In conclusion, these results show that good agreement between E_{stat} (initial conditions) and E_{dyn} (procedural average) can be achieved for polymer decompositions, but only where the mechanism is invariant with extent of reaction. In all other cases, differences between the two sets of data must be expected, unless the dynamic thermograms are processed to allow for variations in mechanism. When the two methods give different results it is likely that those from isothermal data (initial conditions) are more reliable.

ACKNOWLEDGMENTS

We wish to record our appreciation to the Science Research Council for the award of a Research Studentship to one of us, G. P. K., and for the award of a grant to purchase the TGA equipment. We are also grateful to Professor H. Hopff of the Swiss Federal Institute of Technology, Zürich, and to Professor K. Chikanishi of Kyoto University, Japan, for the gift of samples.

REFERENCES

1. J. H. Flynn and L. A. Wall, General treatment of the thermogravimetry of polymers, J. Res. Nat. Bur. Standards U. S.,

70A, 487 (1966).

2. H. Hopff, H. Lüssi, and L. Borla, Zur Polymerisation des Atropasäuremethylesters, Makromolek. Chem., 81, 268 (1965).
3. G. G. Cameron and G. P. Kerr, unpublished results.
4. K. Chikanishi and T. Tsuruta, Reactivity of α-alkylacrylic esters. Part 1. Homopolymerisation behaviors of methyl α-alkylacrylates, Makromolek. Chem., 81, 198 (1965).
5. H. H. G. Jellinek and M. D. Luh, Thermal degradation of isotactic and syndiotactic poly(methyl methacrylate), J. Phys. Chem., 70, 3672 (1966).
6. D. W. Brown and L. A. Wall, Pyrolysis of poly(α-methylstyrene), J. Phys. Chem., 62, 848 (1958).
7. E. S. Freeman and B. Carroll, The application of thermoanalytical techniques to reaction kinetics, J. Phys. Chem., 62, 394 (1958).
8. D. A. Anderson and E. S. Freeman, The kinetics of the thermal degradation of polystyrene and polyethylene, J. Polymer Sci., 54, 253 (1961).
9. L. A. Wall and J. H. Flynn, Degradation of polymers, Rubber Chem. and Technol., 35, 1157 (1962).
10. L. A. Wall, Polymer decomposition: thermodynamics, mechanisms, and energetics, SPE J., 16, 810 (1960).
11. G. G. Cameron and G. P. Kerr, Simultaneous occurrence of chain-end and random initiation during thermal degradation of poly(methyl methacrylate), Makromolek. Chem., 115, 268 (1968).

Apparatus for the Accurate Measurements of Magnetic Susceptibility with the Help of a Vacuum Electrobalance

R. Cini and M. Torrini

Istituto di Chimica Fisica dell'Università
Firenze, Italy

ABSTRACT

Best conditions for magnetic susceptibility measurements of diamagnetic or slightly paramagnetic fluids are stated. It is pointed out that the classical Gouy method is the most favorable.

A simple apparatus consisting essentially of a slightly modified Cahn RH Electrobalance, a permanent magnet, and thermostatic devices, is described. This apparatus can operate under high vacuum or in controlled gas atmosphere from −30 to 150 C with an accuracy about 2×10^{-10} emu even on samples with a total weight greater than 50-60 g. Operating procedures are discussed to minimize experimental errors. The apparatus was used for a new determination of magnetic susceptibility temperature coefficient of water over the temperature range 0-80 C; a comparison between the results obtained and the existing data is reported. Finally, the use of this apparatus is suggested for simultaneous adsorption and magnetic-susceptibility change measurements on powders which possess no paramagnetism.

INTRODUCTION

The improvement in theoretical calculations today allows of the utilization of experimental data of diamagnetism in order to achieve a greater knowledge of molecular structure.[1] Even information on relatively weak intermolecular interactions can be obtained from these data[2] as a valuable complement to the data furnished by other techniques, provided that the magnetic susceptibility measurements are carried out with considerable accuracy (greater than 0.1%).

Several types of apparatus have already been designed in order to measure with great accuracy the magnetic susceptibility of diamagnetic or slightly paramagnetic liquids. However, the scatter between the experimental data on the same substance obtained by various authors often shows that the precision which the authors claim is not reliable.

Among the various well-known methods for measuring the magnetic susceptibility,[3] the Gouy method, in its classical arrangement, is the most rigorous, whether in absolute or in relative measurements on liquids. In fact, in this method the top of the sample is practically outside the magnetic field, and therefore the vessel can be flame-sealed, offering the following advantages which are often neglected:

(1) The magnetic susceptibility even of highly volatile liquids can be measured independently of their vapors and of the atmosphere of the measuring apparatus. This is possible for greater temperature intervals.

(2) The substance being examined can be directly collected in the vessel, where it can be preserved at length under rigorous conditions even after preparation and drastic purification processes.

Furthermore, this method allows the use of dilatometer-shaped vessels, and therefore permits contemporary measurements of magnetic susceptibility and density (as a function of temperature), in order to obtain the mass magnetic susceptibility. This is useful for the measurements on mixtures or on solutions whose density is not available from the tables.

The best experimental conditions occur when compensation vessels are used, and with samples of large cross-sectional area, rather than with high magnetic fields.[4] This requires the use of a notable quantity of material. This last condition offers the advantage of greater guarantee in the preparation and manipulation of the samples, whether they are pure liquids or solutions or mixtures, especially if mixtures are made up of volatile components. Moreover, large samples are less pushed away from the vertical by the forces resulting from field gradients, so that a frequent source of error is avoided.[5] The use of rather weak magnetic fields facilitates the attainment of high reproducibility and constancy of the field.

Therefore, experimental conditions mentioned above require accurate measurements of small weight changes on large samples with a total mass of about 40-60 g.

APPARATUS

The basic component of the apparatus consisted of:

(i) a Cahn RH Electrobalance (modified as specified below) with recorder and vacuum system;

(ii) thermostatic devices for the sample;

(iii) a permanent magnet;

(iv) a thermostatic chamber for the entire apparatus.

Electrobalance

The Cahn RH Electrobalance has a capacity (maximum load on the sample suspension) of 100 g. Maximum sensitivity (0.002 mg) is obtained for measurements of weight changes less than 10 mg. The same sensitivity for greater weight changes is achieved only by manually adjusting calibrated weights on the sample suspension. Obviously, this procedure requires either weighing in air or a very cumbersome sequence of operations in order to restore the vacuum or the controlled atmosphere. Also, much time is required to attenuate sufficiently the vertical component of the oscillations of heavy loads. We overcame these limitations by placing inside the vacuum bottle of the electrobalance a device which was operated externally by remote control.

The device consisted of two pairs of small electromagnets, which allowed calibrated weights to be put on or removed from the sample suspension while the electrobalance was working, without appreciably disturbing the system electrically or mechanically.

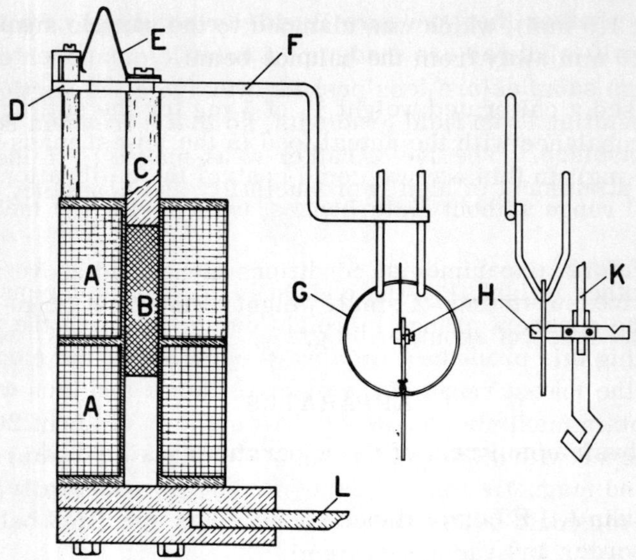

Fig. 1. Weight setting device: A) coils; B) soft iron plunger; C) brass cylinder; D) U-shaped guide; E) friction spring; F) arm; G) calibrated weight; H) weight support; K) suspension ribbon; L) weighing-mechanism base.

This device (Fig. 1) was clamped on base L of the weighing mechanism near the sample suspension. Each pair of electromagnets consisted of a double cylindrical coil A, inside which slid a soft iron plunger B. The plunger was joined to a brass cylinder C, which carried the arm F for the calibrated weight. The two plungers had a vertical stroke of about 15 mm, and they were operated by means of current impulses through the coils. A U-shaped guide D and a friction spring E kept the plunger in the established position, after the releasing current was turned off. The coils were fed through the four free wires of the cable of the electrobalance, as shown by the circuit diagram of Fig. 2. Enameled copper wire of a rather large cross section was employed for the coils in order to expedite the evacuation. The releasing current (1.5–2 A) was supplied by a small variac and rectifier. The brass coils of the electromagnets had a Teflon insulation coating.

The calibrated weights G were aluminum wire rings which, by means of the arms F, were held on a rectangular bar H (cross section 0.5×1.5 mm), which was clamped to the sample suspension strip K, 20 mm away from the balance beam.

We used a calibrated weight P_1 of 5 mg for the setting up of the electrobalance with the actual load in the lowest mass-dial range (10 mg); in this way we could control the calibration of the mass-dial range without disturbances, even during the measurements.

The other weight, P_2, was chosen in order to compensate within 5 mg the apparent change in weight due to action of the magnetic field. Using this procedure we always carried out the measurements in the lowest range of the electrobalance and thus we were able to obtain maximum accuracy. A weight P_2 of about 20–25 mg was employed with diamagnetics ($\chi = 0.6–0.8 \times 10^{-6}$ emu) and with vessels and magnetic fields as specified below. Generally, weights up to 200 mg could be used.

Since the vessel is never perfectly cylindrical, it might rotate under the action of the magnetic field. To reduce these rotations the length of the suspension ribbon was reduced to 30 mm, and a copper flat hook was sealed to the strip. The vessel was hung up to this hook by rigid pyrex or silica fibers which were provided with terminal acute-angle-shaped hooks in order to hinder any rotation.[5]

The electrostatic troubles were eliminated by means of a small radioactive source.

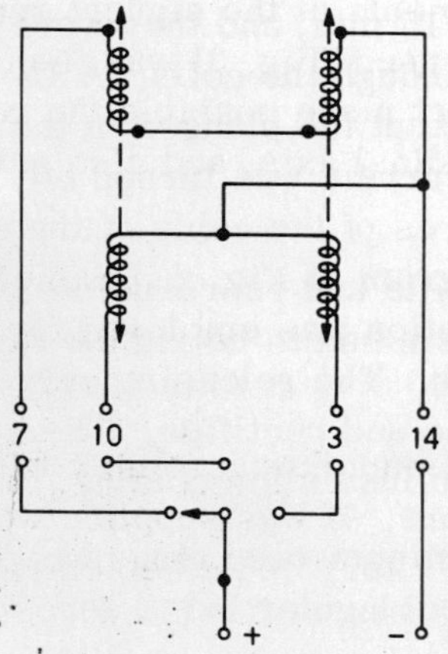

Fig. 2. Feeding circuit of the weight-setting device. 7, 10, 3, and 14 are the numbers on the plugs corresponding to the four free wires of the cable from the control unit to the weighing unit.

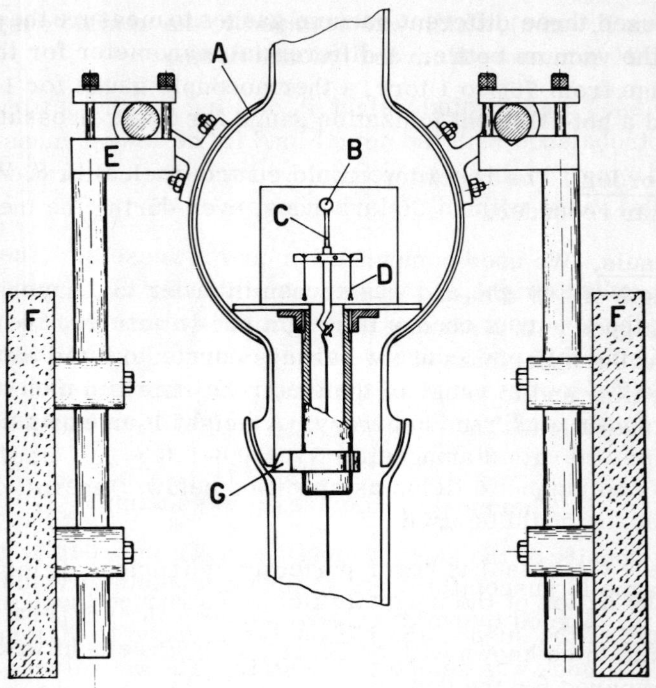

Fig. 3. Support of the weighing unit: A) vacuum bottle; B) balance beam; C) suspension ribbon; D) calibrated weight; E) brass support; F) marble wall brackets; G) centering block.

Weighing-Unit Assembly. A very rigid support for the weighing mechanism is required for experiments at the highest sensitivity with heavy loads.[7] The support we used (Fig. 3) was fixed to two marble wall brackets. This support made possible the rotation of the vacuum bottle around its longitudinal axis, and also gave it the freedom needed for adjustment.

In order to give the sample a definite and reproducible position in the field, the weighing mechanism had to be rigidly fastened to the vacuum bottle.

Vacuum Pump. We used a Galileo pumping assembly with a capacity of 300 liters $\cdot$ sec^{-1} at 10^{-6} torr. It was supplied with a gas-inlet microvalve. A vacuum of 10^{-7} torr could be achieved by using a liquid air trap.

We used three different vacuum gauges to measure the pressure in the vacuum bottle. A differential manometer for the pressure range from 760 to 1 torr, a thermocouple gauge for 1 to 10^{-3} torr, and a hot-filament ionization gauge for lower pressures.

Recorder. The recorder employed was an L & N. S. W. Speedomax Azar recorder.

Vessels. We used compensation quartz vessels.[4] They had a total length of 400 mm and an internal diameter of 11 mm. They were designed with a narrow neck (2 mm in diameter, 80 mm in length) at the upper part of the vessel in order to facilitate the flame sealing and to avoid a significant decomposition of vapor during this operation. The neck was provided with an expansion bulb.

Thermostatic Devices for the Sample

The density of the sample must be known in order to get the mass magnetic susceptibility from measurements of volume susceptibility. A good thermoregulation enables the temperature of the sample to be known with an accuracy in agreement with the precision required for the density.

The temperature of the sample was controlled by means of a high-delivery rapid flow of liquid through a double-walled tube[6] which surrounded the sample vessel. It was verified that temperature differences greater than 0.01 C did not exist between the inlet and the outlet of the tube. The thermal gradient over the whole length of the sample was then less than 0.005 C. The periodic temperature fluctuations in the thermostatic flow were reduced to ±0.01 C by employing two thermostats, while the mean temperature was constant within ±0.002 C. One of these controlled the temperature of the liquid flowing around the sample; the second supplied the cooling flow to the first. For low temperatures the second thermostat was replaced by a cryostat.

The actual temperature of the sample, which was read at the inlet of the double-walled tube, was known better than ±0.05 C.

The double-walled tube was jointed to the hangdown tube of the electrobalance by means of a spherical joint. Too low a pressure had to be avoided in order to obtain a rapid thermal exchange between the wall of the thermostatic tube and the sample. Therefore, the measurements were carried out in an atmosphere of dry N_2 under a reduced pressure which was above the pressure where the thermomolecular flow took place and below the pressure where convective motions became harmful (50 torr). Here the correction due to the volume magnetic susceptibility of N_2 surrounding the vessel was negligible. Two annular supports were placed in the upper part of the thermostatic tube; the vessel suspension put two mica disks on these supports when the sample was accommodated inside the tube.[6] These disks guaranteed the uniformity of the temperature and prevented the heating of the apparatus at high temperatures as well as any trouble due to convection motions even at pressures of 100-200 torr.

The thermostatic tube was supplied with a very good thermal insulating coating (asbestos and expanded polystyrene). The apparatus can be used in the temperature range from −30 C to +150 C.

Permanent Magnet

The disadvantages which generally occur by using permanent magnets instead of electromagnets do not exist for magnetic measurements on liquids. Therefore we preferred to employ a permanent magnet because of the simplicity in its use.

The Alnico magnet built by the Indiana Steel Products Co. produced a field of about 9000 Oe in an air gap width of 2.5 cm between the pole caps, which were 5 cm in diameter. The uniformity of the field was sufficient with respect to the reproducibility of the position of the sample and of the magnet within the limits of the precision of the electrobalance.

The magnet was mounted on a carriage on ball bearings, which was moved by a servomotor on a rectified horizontal guide. The ball bearings were mounted in such a manner as to determine a three-point supporting system for the carriage. The orientation of these ball bearings rigorously restricted the movement in a plane along an axis (Fig. 4). The sharply defined stopping point made possible the applying and removal of the magnet with a reproducibility of position of 0.005 mm, and therefore also of the field around the specimen within the limits of reproducibility required for the measurements.

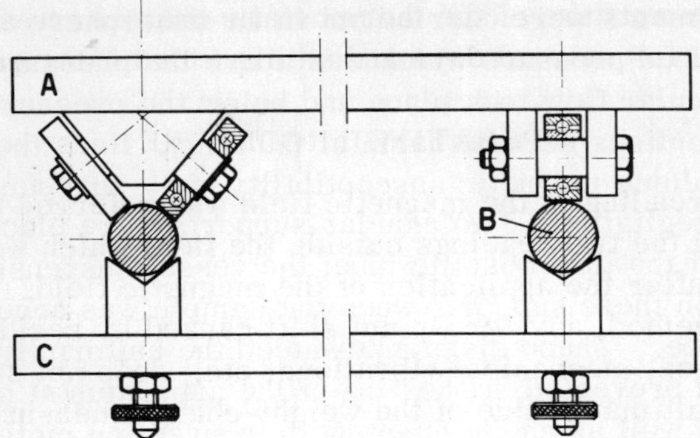

Fig. 4. Magnet guide assembly: A) magnet carriage; B) rectified steel bar; C) supporting frame.

There was a distance of 90 cm between the magnet and the electrobalance; the polar axis of the magnet was perpendicular to the axis of the weighing-mechanism magnet. No appreciable disturbance was noticed for such a distance with that orientation when moving the magnet along the entire stroke of the slide. This was confirmed by measurements made by replacing the sample with a weight which was identical except that it was hung on a notably shortened suspension.

Thermostatic Chamber

In order to have a magnetic field which was sufficiently constant, the temperature of the magnet had to be maintained within ±0.1 C, and the electrobalance required a constancy of temperature within ±0.05 C in order to have high zero-point stability with heavy loads at the maximum sensitivity for the entire duration of the measurements.[7] Thus the weighing unit and the magnet assembly were placed in a thermostatic room.

The dynamic air flow[8] was used in order to control the temperature to within ±0.05 C, and two thermocouples were put on the pole caps for temperature control. Dynamic and not static thermoregulation was used to ensure greater temperature stability of the magnet.

The thermostatic chamber had walls of expanded polystyrene (50 mm thickness) and its capacity was 9 m^3.

The temperature of the thermostatic chamber was constant at 20.0 ± 0.1 C for several days and within ± 0.05 C for an hour.

OPERATING PROCEDURE

Every reading in the magnetic field was referred to the average between the two readings outside the field which were taken before and after the application of the magnetic field. With this operating method, any zero-point shift caused by residual electrostatic charges, mechanical vibrations, etc., does not significantly impair the ultimate value of the weight-change measurement.

RESULTS

As a first application of the apparatus described we undertook a new determination of the magnetic susceptibility of water as a function of temperature. The results of this investigation will appear in another publication.[9] This research offered the opportunity for a rigorous check of the apparatus.

The measurements were carried out according to the operating procedure described above. Five measurements of apparent weight change were made at a given temperature. The maximum deviation which was found between two readings of the same set was 0.005 mg, whereas the mean deviation from the average was about 0.001 mg for each set of measurements. Before and after the readings which were carried out at a given temperature a set of reference measurements at 20° was taken. The average values of these reference measurements practically coincided within an interval of several hours.

The estimated error on the ratio χ/χ_{20} measurements was about 1.6 per 10,000 on three different specimens of water.

The curve derived from our measurements by the least-squares method[9] is compared with the most reliable data of other authors[10] in Fig. 5.

These authors do not take into account particular cautions in order to avoid thermal gradients along their specimens. It is expedient to note that these gradients would have been very harmful in Auer's, Seely's, and Wills and Boeker's experiments, which were carried out by the Quinke method. We think that the presence of such gradients can be one of the causes of the disagreement be-

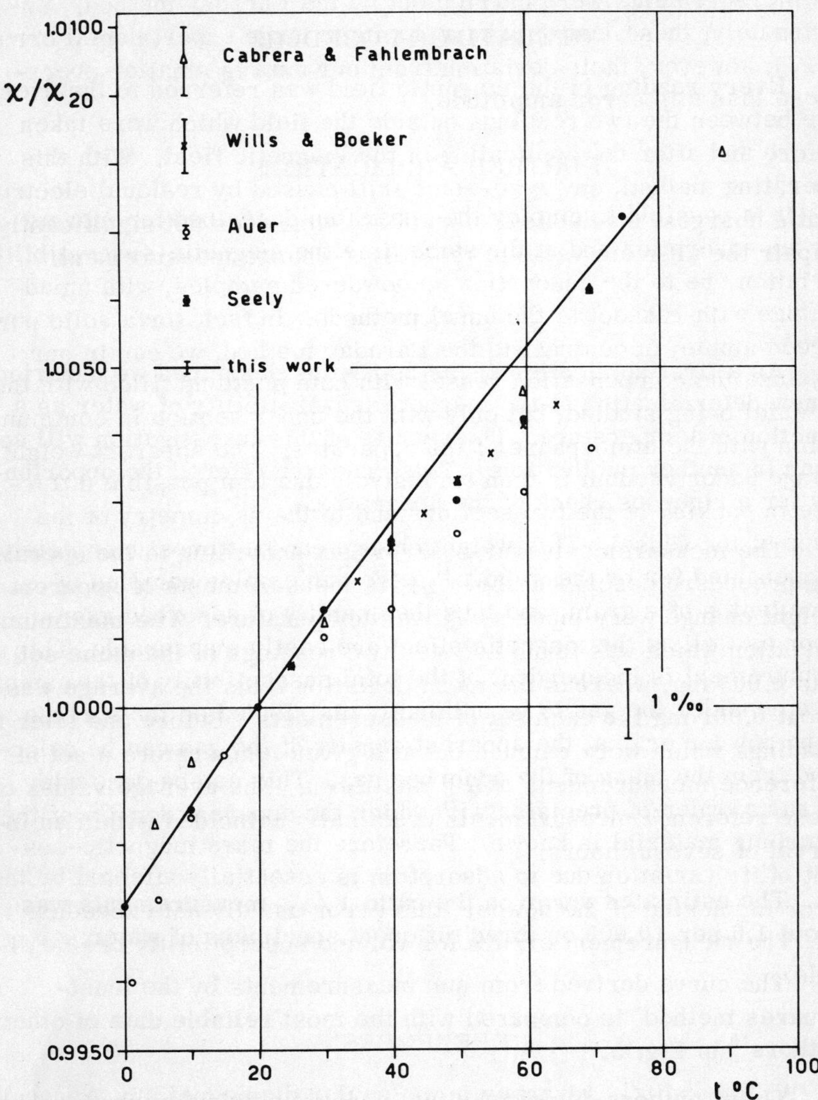

Fig. 5. Variation of the magnetic susceptibility of water with temperature. Vertical segments refer to the experimental error given by each author.[9,10]

tween their data and ours. On the other hand, the data of Cabrera and Fahlembrach were insensitive to thermal gradients, because the measurements were carried out by the Faraday method. Unfortunately, these last data have a rather large experimental error (0.1%); however, their deviation from our data is smaller everywhere than this error amplitude.

FURTHER APPLICATIONS

It is possible to employ the apparatus described for measuring the adsorption and at the same time the magnetic-susceptibility variation due to the adsorption on powdered samples, with an advantage with respect to the usual methods. In fact, for a solid powdered sample, in contrast to the Faraday method, we can in our case use the compensation vessel with both sections filled with the material being studied, but only with the upper section in communication with the atmosphere of the apparatus. The apparent weight change under vacuum is then exclusively due to a possible difference in packing of the two sections and to the asymmetry of the shape of the vessel. This weight change can be almost completely compensated for by the weight P_2. We can employ samples of some tenths of a gram, and thus the quantity of adsorbed gas or vapor as well as the magnetic effect are easily measurable. This measurement is independent of the total susceptibility of the sample. If we consider the gas to be uniformly distributed on the surface of the powder grains, the apparent density of the gas can be calculated from the mass of the adsorbed gas. This can be done with the same order of precision with which the apparent density of the adsorbing material is known. Therefore the mass magnetic-susceptibility variation due to adsorption is essentially affected by the error in packing of the powder, this error usually not exceeding 3%. The measurement errors for volume susceptibility are negligible.

REFERENCES

1. W. Haberditzl, Advances in molecular diamagnetism, Angew. Chem., Intern. Ed. (Engl.), 5, 288 (1966).
2. Ya. G. Dorfman, Diamagnetism and the Chemical Bond, Edward Arnold Ltd., London (1965), pp. 46-68.
3. P. W. Selwood, Magnetochemistry, Interscience Publishers, New York (1956), pp. 3-19.

4. L. Michaelis, Determination of magnetic susceptibility, in: Physical Methods of Organic Chemistry, Vol. 2 (A. Weissberger, ed.), Interscience Publishers, New York (1946), pp. 1220, 1223.
5. R. Cini, Su di alcuni dispositivi sperimentali per ricerche magnetochimiche. Nota III. Apparecchiatura per la misura della suscettività magnetica di massa col metodo di Faraday, Ricerca Sci., 29, 772 (1959).
6. R. Cini, Su di alcuni dispositivi sperimentali per ricerche magnetochimiche. Nota I, Ricerca Sci., 29, 272 (1959).
7. W. C. Tripp, R. W. Vest, and N. M. Tallan, System for measuring microgram weight changes under controlled oxygen partial pressure to 1800°C, in: Vacuum Microbalance Techniques, Vol. 4 (P. M. Waters, ed.), Plenum Press, New York (1965), p. 146.
8. R. Cini and L. Sacconi, An apparatus for the accurate measurement of magnetic susceptibility, J. Sci. Instr., 31, 56 (1954).
9. R. Cini and M. Torrini, Temperature dependence of the magnetic susceptibility of water, J. Chem. Phys. 49, 2826 (1968).
10. P. W. Selwood, Magnetochemistry, Interscience Publishers, New York (1956), pp. 86-87.

Reduction Characteristics of Certain Oxides of Nickel and Uranium

P. S. Clough and D. Dollimore

The Chemistry Department
The University of Salford
Salford, Lancs., England

ABSTRACT

The reduction of a uranium oxide (U_3O_8) and nickelous oxide (NiO) by hydrogen atmospheres has been studied using a Cahn RG electrobalance. The oxide samples were prepared by the thermal decomposition of selected nickel and uranyl salts. Heating of the product oxides caused sintering observable as a decrease in surface area. The effect of this sintering on the rates of reduction of the oxides has been investigated in a series of isothermal experiments. The kinetics of the reduction process are investigated and possible reaction mechanisms are suggested.

INTRODUCTION

The uranium — oxygen system is one of the most complex oxide systems known, owing in part to the multiplicity of oxidation states of comparable stability.[1] It is well known that reduction of the higher oxides of uranium with hydrogen leads to the formation of $UO_{2.0}$. Perhaps the first to describe this process was Arfredson in 1822.[2] Since this date the system has been quite widely studied due to its importance in the refining of uranium ores. It is only relatively recently that agreement has been reached as to the conditions required for complete reduction to $UO_{2.00}$. As late as 1943 it was asserted that at temperatures of 650° the reduction stopped at the composition $UO_{2.14}$. More recent work, however,[3,4] indicates that stoichiometric UO_2 is produced at lower temperatures. It has been shown that the rate of reduction of U_3O_8 depends to some extent on the nature of its preparation,[3] although in a recent study Dell and Wheeler,[4] using samples of very low surface area U_3O_8 prepared over a wide range of temperature found the reduction rate to be quite constant, although they also report that reoxidation leads to the production of a much higher surface area U_3O_8, which reduces at a faster rate than the original samples. The rates of reduction obtained by the latter authors were observed to be almost linear over a large extent of the reaction.

De Marco and Mendel[5] studied the reduction of high-surface-area UO_3, finding it to take place in two stages involving the formation of an intermediate $UO_{2.56}$. Notz and Mendel,[6] using relatively low surface area UO_3 suggested a three-stage process:

$$UO_3 \rightarrow U_3O_{8+} \rightarrow U_3O_{8-} \rightarrow UO_2.$$

They demonstrated that the rate for this reduction process is directly proportional to the surface area of the starting material.

In view of the industrial importance of supported nickel—metal catalysts, it is perhaps surprising that the reduction of nickel oxide, which is a necessary step in their preparation, has not been more widely studied. Early work in this field has been reviewed by Mellor.[7] The reduction has been reported to begin at temperatures as low as 120 C,[8] but this is not supported by the evidence of other studies,[9-11] where temperatures in excess of 150 C have been found necessary before the reaction commences. Parravano,[9] and Iida and Shimada,[10] studied the effect of additives on this reduction process. More recently Delmon[11] found the process to follow a power-law kinetic expression for the acceleratory period of the reaction. Vlasenko and Telipko[12] observed that the reaction had an induction period the length of which varied with temperature; during this induction period no perceptable weight change took place. The present study was undertaken to establish the most favorable conditions for the reduction of the various oxide samples to nickel metal and uranium dioxide, the active catalytic species in certain petroleum reforming processes.

SAMPLES

The oxide samples used in this study were prepared from the respective acetates and nitrates by their thermal decomposition in air at the temperatures indicated in the sample designation. For instance, sample UA_{300} was prepared from uranyl acetate by decomposition at 300 C.

APPARATUS

The reduction processes were carried out on a Cahn RG electrobalance. The sample weights were of the order of 50 mg, while the weight loss on reduction was of the order of 2 mg for the uranium oxides and 10 mg for the nickel oxides. Care was taken to spread the samples evenly in a thin layer, since as similar packing conditions would apply when smaller sample weights were used, it was considered unlikely that use of smaller sample weights would have any effect on the ultimate accuracy of the results. This was shown to be the case by initial experimentation.

The furnace used was noninductively wound, and sealed at one end. It was placed around the sample tube prior to an experimental run and brought to the required temperature using a variac controller, which could also be controlled by a motor-driven cam to obtain linear heating rates in rising-temperature experiments. The balance was shielded from furnace radiation by wrapping asbestos string around the sample tube, covering the slight gap between the tube and furnace. Aluminum foil was wrapped around the sample tube to about 20 cm above the top of the furnace, and an asbestos sheet was fitted immediately above the balance-case quick-fit joints. The balance case itself was wrapped with aluminum foil to prevent any stray radiation entering the photocell compartment.

Sample temperature measurement was effected by means of a Cahn type 2020 thermocouple assembly; the hot junction was suspended in close proximity to the sample, while the cold junction was suspended in the hang-down tube, its tip immersed in Apiezon B oil, and was maintained at 0 C. Use of the 2020 thermocouple assembly served not only as a means of accurate sample-temperature assessment, but also to reduce buoyancy effects to negligible proportions, as shown by heating 50-mg samples of dead-burnt alumina from ambient temperatures to 500 C in hydrogen atmos-

Table I

Sample designation	Surface area, $m^2 \cdot g^{-1}$	Activation energy, $kcal \cdot mole^{-1}$
UA_{300}	7.9	37.1
UA_{600}	5.0	35.1
UA_{900}	2.5	33.6
UN_{650}	1.1	34.2
NA_{350}	22.5	28.0
NN_{500}	14.6	23.0

pheres. Water formed as a reduction product was trapped out using liquid nitrogen.

RESULTS

Reduction of U_3O_8

Details of the specific surface area of the samples used are given in Table I. Uranyl acetate had previously been shown[13] to decompose quite rapidly to produce U_3O_8 samples composed of small crystallites of $\sim 10^{-1}\ \mu$ in size. The U_3O_8 thus produced was shown to have a relatively high surface area which decreased quite rapidly with temperature; these samples were also shown to be nonporous in nature. In the light of this information, samples of U_3O_8 were prepared by decomposition of uranyl acetate in air at 300 C, the samples being removed from the furnace soon after the decomposition was complete, as detected by weight loss. This sample, as previously stated, was referred to as UA_{300}. Further samples were prepared by heating sample UA_{300} at temperatures of 600 C and 900 C for a short period of time sufficient to produce a series of samples of diminished surface area. The fourth sample in the series, UN_{650}, was prepared by slow decomposition of uranyl nitrate at 300 C to produce UO_3, then by placing the oxide so produced in a furnace at 650 C for a short length of time to effect the conversion to U_3O_8.

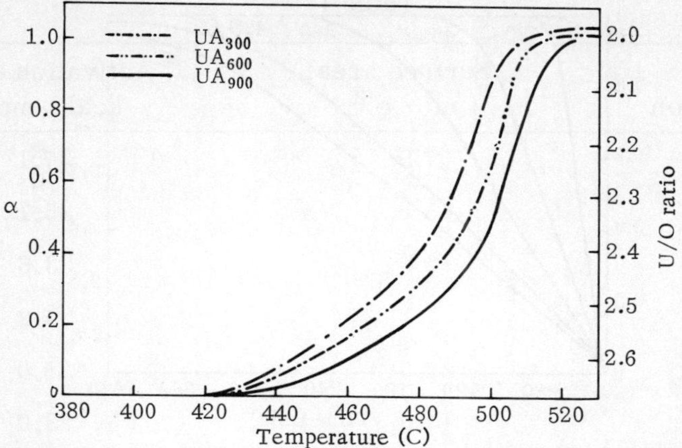

Fig. 1. Thermograms of reduction of U_3O_8 samples with hydrogen.

Figure 1 gives details of the thermogravimetric records obtained for the reduction of U_3O_8 to UO_2. The heating rates used for these runs was 2.5 C $\cdot$ min^{-1}. On the basis that the composition of the starting material corresponded to the formula $UO_{2.667}$ (i.e., U_3O_8) the recorded weight loss corresponded to the formula $UO_{1.98}$. It was thought probable that this excess weight loss was due to a slight excess of oxygen in the U_3O_8 lattice, as reduction beyond the stoichiometry $UO_{2.00}$ does not take place in hydrogen atmospheres. Figure 1 also shows that the reduction process for the samples

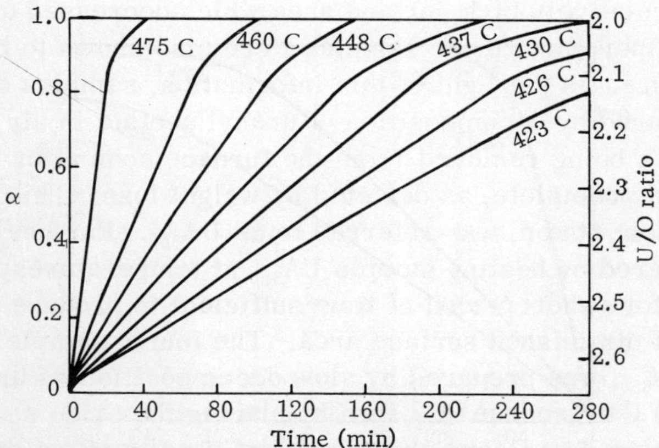

Fig. 2. The isothermal reduction of sample UA_{300}; experimental temperatures as indicated.

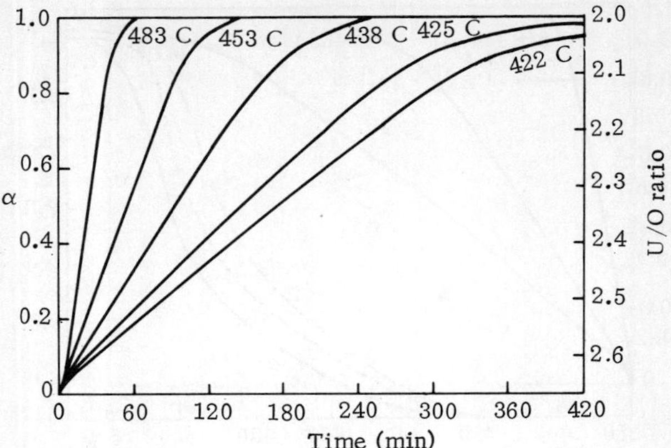

Fig. 3. The isothermal reduction of sample UA_{600}; experimental temperatures as indicated.

studied begins in the range 420 to 430 C, and is complete by 510 to 530 C depending on the nature of the sample.

Isothermal reduction of sample UA_{300} yielded the data shown in Fig. 2. These plots are clearly almost linear over a wide range of the reduction process, only diverging from linearity after about 75% reduction, which corresponds to the stoichiometry $UO_{2.15}$. Figures 3, 4, and 5 show data obtained for the reduction of samples UA_{600}, UA_{900}, and UN_{650}, respectively. These plots can also be seen to be linear over a similar range of the reduction.

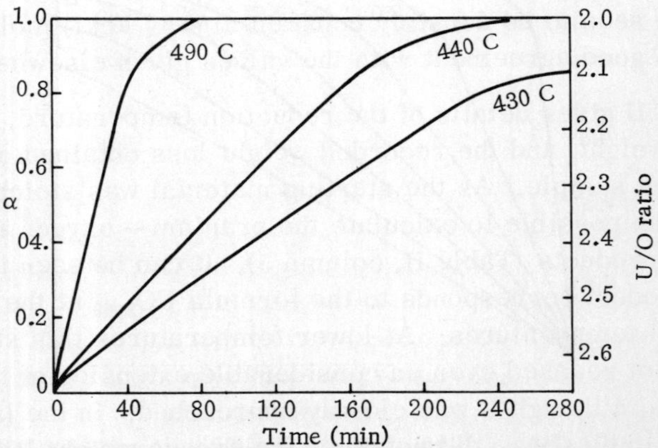

Fig. 4. The isothermal reduction of sample UA_{900}. Experimental temperatures as indicated.

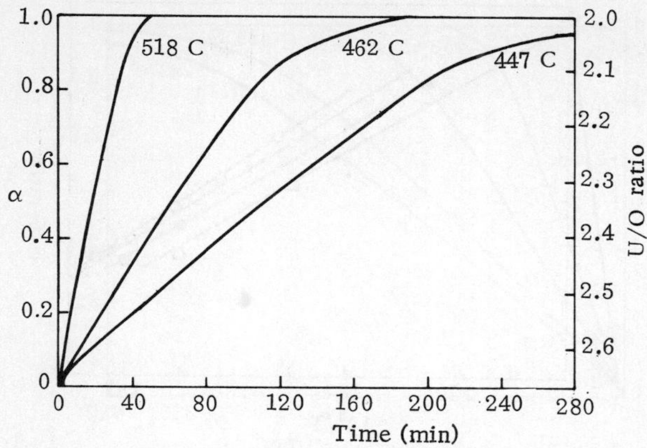

Fig. 5. The isothermal reduction of sample UN_{650}; experimental temperatures as indicated.

It was possible to make a direct assessment of the rate constants for the reduction of the various samples due to the linear nature of these processes. The activation energies for the linear region of the reduction process for each sample was calculated using the Arrhenius equation, i.e.,

$$k = A\exp(-E/RT)$$

where k is the rate constant, E is the activation energy, R is the gas constant, and T is the isothermal temperature in degrees K. A is a constant often called the preexponential term. Plots of log k versus 1/T are linear (Fig. 6), and the values of the activation energy calculated from the slopes of these lines are given in Table I. These are seen to be sensibly constant at ~34 kcal/mole, which is in quite good agreement with the values given elsewhere.[3,4,6]

Table II gives details of the reduction temperature, the initial and final weight, and the recorded weight loss obtained on reduction of each sample. As the starting material was stoichiometric U_3O_8, it was possible to calculate the uranium — oxygen ratio in the reduction products (Table II, column 5). It can be seen that the reduction product corresponds to the formula $UO_{2.00}$ at the higher experimental temperatures. At lower temperatures this stoichiometry was not reached even on considerable extension of the experimental run, although it was closely approached. In the isothermal runs the sample UA_{300} did not show the excess weight loss previously associated with excess oxygen in the U_3O_8 lattice in the rising

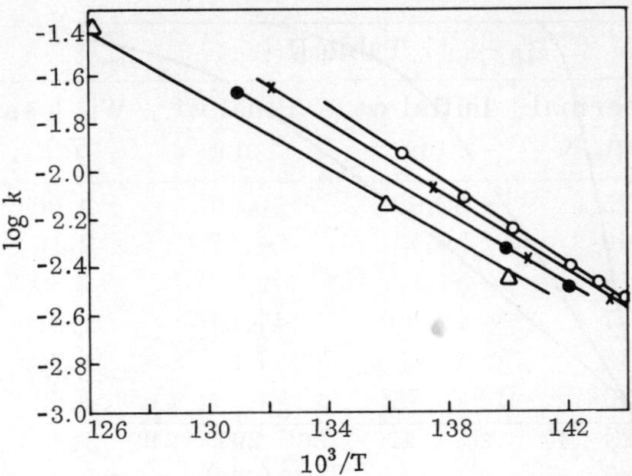

Fig. 6. Arrhenius plots: ○) UA_{300}; ×) UA_{600}; ●) UA_{900}; △) UN_{650}.

temperature experiment (Fig. 1). This was due to the fact that the sample had been evacuated and maintained at temperatures up to 400 C prior to reduction run. In fact it was found that the pretreatment of sample UA_{300} caused a degree of sintering to take place resulting in a surface area of 7.9 $m^2 \cdot g^{-1}$, as opposed to the original 14.8 $m^2 \cdot g^{-1}$.

Using the Arrhenius plots shown in Fig. 6, it was possible to obtain values for the rate constant for the series of samples at a particular temperature although these values were not obtained experimentally, they serve to demonstrate the relationship between surface area and reduction rate, which was as shown in Fig. 7. This figure shows that for the samples studied a linear relationship existed between these two parameters; however, since there must be zero rate at zero surface, there must be a sudden drop in rate with lower values of surface area.

Reduction of Nickelous Oxide (NiO)

Table I gives details of the two nickel-oxide samples investigated in this study. Sample NN_{500} had been prepared by the thermal decomposition of nickel nitrate hexahydrate in air at 500 C. Thermogravimetric studies had previously shown the decomposition residues to contain occluded oxides of nitrogen at temperatures much below 500 C. Isothermal adsorption of nitrogen at 77 C revealed this sample to be nonporous in nature. Sample NA_{350} had been prepared from nickel acetate by decomposition in air at 350 C. Leic-

Table II

Sample	Isothermal temp., C	Initial wt., mg	Final wt., mg	Wt. loss, mg	U/O ratio in product
UA_{300}	475	52.69	50.69	2.00	2.00
	460	56.73	54.57	2.16	2.00
	448	51.12	49.23	1.89	2.02
	437	49.00	47.19	1.81	2.02
	430	49.68	47.89	1.78	2.05
	426	48.86	47.24	1.62	2.08
	423	56.26	54.49	1.77	2.10
UA_{600}	485	58.68	56.45	2.23	2.00
	453	57.12	54.95	2.17	2.00
	438	56.03	54.00	2.03	2.03
	425	46.12	44.44	1.68	2.02
	422	49.86	48.21	1.65	2.08
UA_{900}	490	60.08	57.80	2.28	2.00
	440	54.62	52.40	2.08	2.00
	432	54.33	52.46	1.87	2.06
UN_{650}	518	51.90	49.95	1.95	2.00
	462	52.66	50.68	1.98	2.00
	447	50.13	48.27	1.86	2.02

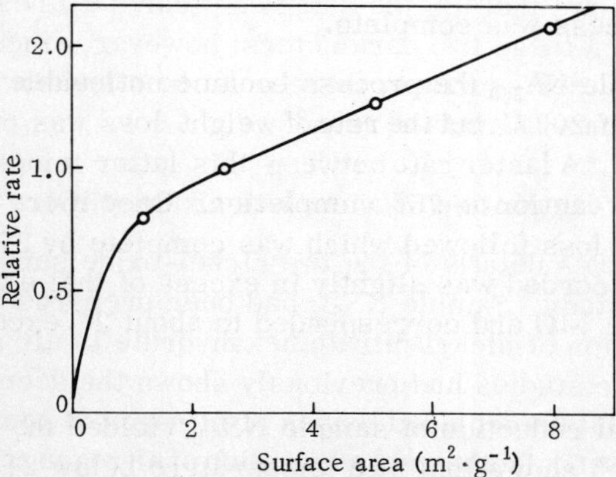

Fig. 7. Change of reduction rate with surface area.

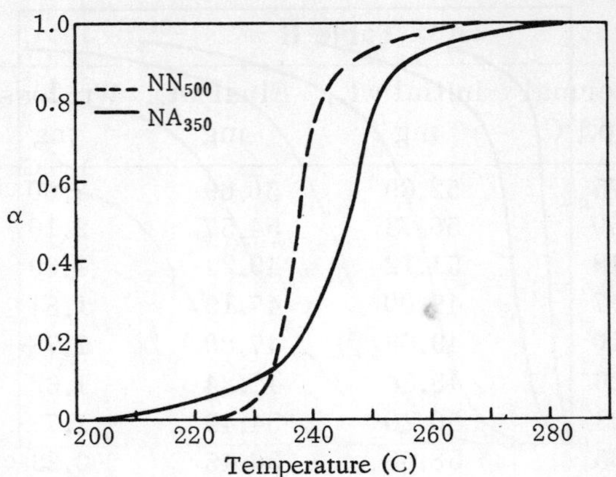

Fig. 8. Thermogram of reduction of NiO samples with hydrogen.

ester and Redman[14] had previously investigated the solid residues from this process and thought it probable that they would be porous in nature. Nitrogen adsorption at 77 C revealed this to be so, the residues having a narrow range of pore sizes ~100 Å.

Figure 8 shows details obtained on reduction of these two samples under conditions of linearly increasing temperature (1.5 C · min^{-1}). They show that for sample NN_{500} the reduction process became noticeable at 220 C; the rate increased slowly, at first becoming faster until, at 240 C, 80% of the starting material had been reduced. At this point a period of slow weight loss began, and at 270 C the process was complete.

For sample NA_{350} the process became noticeable at the lower temperature of 200 C, but the rate of weight loss was only slow until about 230 C. A faster rate between this latter temperature and 250 C took the reaction to 75% completion. Once more a period of slower weight loss followed which was complete by 270 C. The final weight loss recorded was slightly in excess of that expected for stoichiometric NiO and corresponded to about 2% excess oxygen in the lattice.

Isothermal reduction of sample NN_{500} yielded the data shown in Fig. 9, which shows that at a temperature below 215 C the reduction had an induction period where there was no detectable change in weight. This induction period can be seen to vary in length from 5 min at 213 C to 100 min at 180 C (Fig. 10). This figure would seem

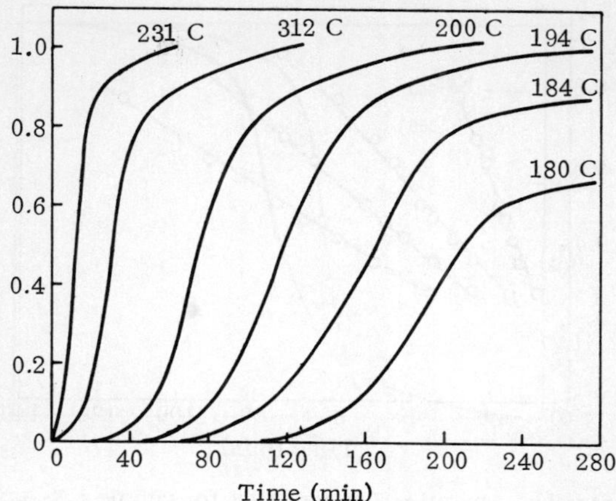

Fig. 9. Isothermal reduction of sample NN_{500}.

to bear out the experimentally observed fact that at 175 C the reduction process did not begin even after 6 hours. A kinetic analysis of the isothermal data revealed that between $\alpha = 0.025$ and $\alpha = 0.75$ the process followed the Avrami – Erofeyev equation, which can be modified to take the form

$$\log\log(1/1-\alpha) = n\log t + \text{const} \tag{1}$$

or

$$\log(1/1-\alpha) = kt^n \tag{2}$$

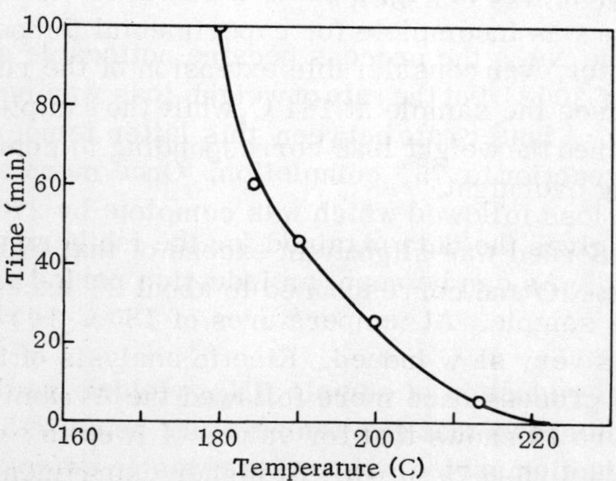

Fig. 10. Length of induction period versus temperature for sample NN_{500}.

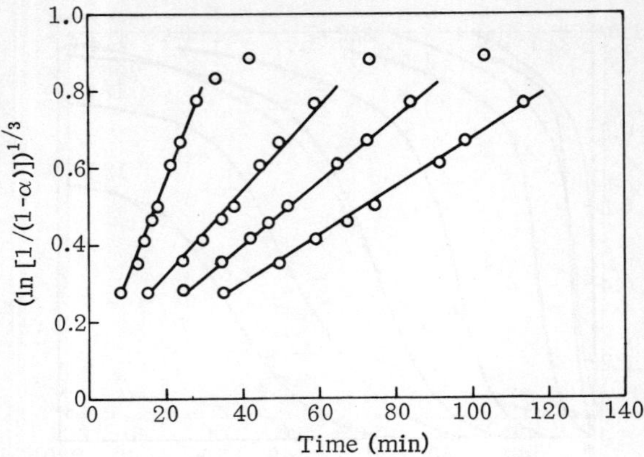

Fig. 11. Avrami — Erofeyev plots for sample NN_{500}.

where α is the fraction of reactant that has reacted after time t, and k is the rate constant.

Thus, by using Eq. (1), it was possible to calculate a value for n from the slope of the line obtained by making plots of log log $[1/(1-\alpha)]$ versus log t. Using the values of n thus obtained, values for k, the rate constant, were calculated from plots of $(\log[1/(1-\alpha)])^n$ versus t (Fig. 11). The Avrami — Erofeyev equation was developed to describe reactions in which overlap of growing nuclei occurs, so that it was considered probable that this process was followed by the reaction under consideration. Further consideration of the isothermal reduction data in Fig. 9 showed that after 240 min the reduction process was incomplete for experimental temperatures below 200 C. In fact, even considerable extension of the run failed to completely reduce the sample at 184 C, while the sample reduced at 190 C only reached the weight loss corresponding to complete reduction on standing overnight.

Figure 12 gives the data obtained for the isothermal reduction of sample NA_{350}. As can be seen, an induction period was not observed for this sample. At temperatures of 180 C the rection commenced but was very slow indeed. Kinetic analysis of this data revealed that the process once more followed the Avrami — Erofeyev equation. Figure 13 shows that for values of α up to ~0.25, n = 1, whereas for α = 0.25-0.75, n = 3. At higher experimental temperatures an excess weight loss was once more noticed, again corre-

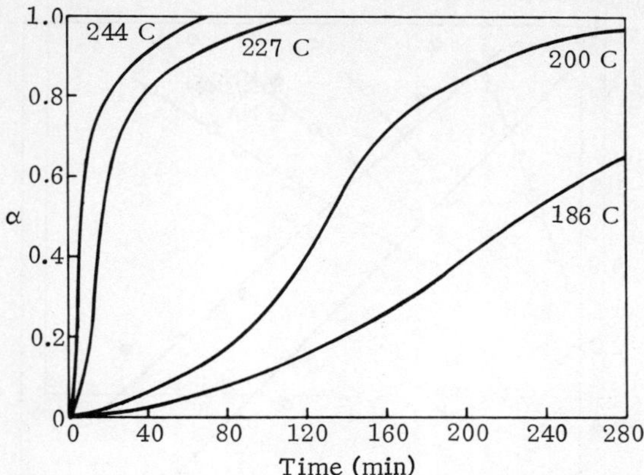

Fig. 12. Isothermal reduction of sample NA_{350}; reduction temperatures as indicated.

sponding to about 2% excess oxygen. The percentage weight loss in the experiments at the lower temperatures is expressed as a percentage of this total experimental weight loss, and not as a percentage of the weight loss corresponding to stoichiometric NiO.

Values of the rate constants obtained from the kinetic plots as previously described when used in conjunction with the Arrhenius equation and plotted in the usual form of log k versus 1/T, gave

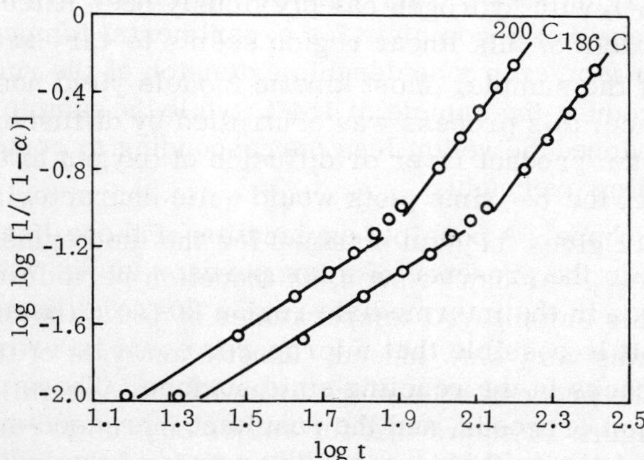

Fig. 13. Avrami – Erofeyev plots for sample NA_{350}; experimental temperatures as indicated.

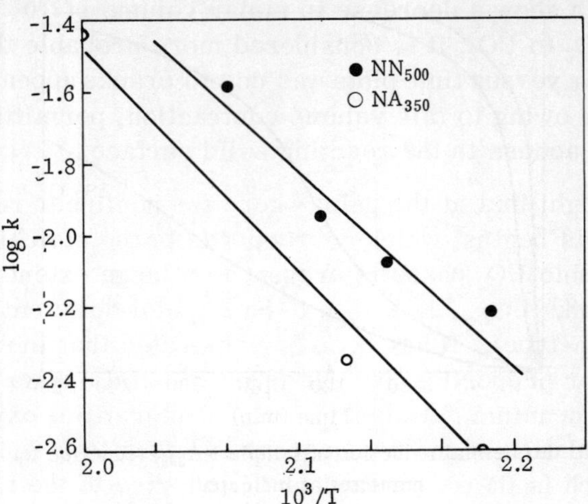

Fig. 14. Arthenius plots for NiO samples

values of 23 kcal · mole^{-1} for the reduction of sample NN_{500}, and 28 kcal · mole^{-1} for sample NA_{350} (Fig. 14).

DISCUSSION AND CONCLUSIONS

Reduction of U_3O_8

The approximately linear plots given by the isothermal reduction of U_3O_8 with hydrogen has previously been noted,[4,6] although the extent of this linear region seems to vary according to the nature of the sample. Most kinetic models yield nonlinear data. If, for instance, this process was controlled by diffusion of hydrogen through the product layer or diffusion of oxygen to the reacting solid surface, the α — time plots would quite characteristically be parabolic in shape. A possible explanation of these linear rates could be due to the presence of micropores ~ 20 Å. While the absence of pores in the intermediate region 20–200 Å has been demonstrated,[13] it is possible that micropores could be present as fissures and cracks in the reacting solid surface. These pores widen as the reaction proceeds, and the resultant increased surface could theoretically balance the decrease in surface of the bulk solid, so that an almost constant reaction interface is presented up to quite high conversion ratios. However, on the basis of density determinations, which show a decrease in molar volume of 30% on transition from U_3O_8 to UO_2, it is considered more probable that the linearity of the α versus time plots was due to cracks appearing in the product layer owing to this volume contraction, permitting constant unrestricted access to the reacting solid surface.

It is thought that at the point where the nonlinear region in the reduction plots begins, which corresponds to the stoichiometry $UO_{2.15}$, the cubic UO_2 phase is present to a large extent in a highly defective form. UO_{2+x} is known to be a metal deficient semiconductor (i.e., p-type). It has been demonstrated that in fact these semiconductor properties are due to interstitial oxygen ions, as opposed to a uranium deficiency.[16] Each interstitial oxygen ion induces the oxidation of the two nearest U^{4+} ions to U^{5+}. The mechanism which is therefore thought to give rise to the nonlinear region beginning at ~75% conversion is dependent on the formation of the cubic UO_{2+x} phase in quantities significant enough to influence the kinetics of the reduction process.

If the mechanism is one of initial adsorption of hydrogen at these defect-surface cationic sites according to the equation

$$H_2 + U^{5+}/\square_S^- \between \square^- \rightleftharpoons H_2^+/\square_S^+ + U^{4+}/\square^+ \tag{2}$$

then these chemisorbed H_2^+ ions can react with neighboring oxide ions, at or near to the surface:

$$H_2^+/\square_S + O^{2-}/\square^- \rightleftharpoons H_2O\uparrow + \square_S^+ + \square^-/e \tag{3}$$

The electrons formed by this process will migrate into the crystal interior to reduce internal defect cationic sites,

$$\square^-/e + U^{5+}/\square^+ \rightleftharpoons U^{4+}/\square^+ + \square^- \tag{4}$$

leaving a vacant anionic surface site which will be filled by an interstitially-held oxide ion:

$$O^{2-}/\Delta^- + \square^- \rightleftharpoons O^{2-}/\square^- \tag{5}$$

It was realized that there would probably be a considerable contribution to the reduction mechanism by diffusion of the H_2^+ ions of Eq. (2) into the lattice interior, or by formation of hydroxyl ions as an alternative to Eq. (3); both these reactions would lead to a regeneration of the original surface cationic defect site. The diffusion process postulated in Eq. (5) will depend to a great extent on the size of the crystallites and on the ambient temperature, as it is an activated process. This could explain the almost asymptotic approach toward the stoichiometry of $UO_{2.00}$ observed at the lower experimental temperatures.

The data in Figs. 1-4 would appear to establish that the reduction of U_3O_8 with hydrogen depends to a great extent on the surface area or size of the U_3O_8 particles. The probability of this relationship had previously been pointed out by Harrington and Ruehle[3] and confirmed to some extent by Dell and Wheeler,[4] who demonstrated that particles that had been fragmented by oxidation — reduction cycles reduced at a much faster rate than the original material. The dependency of the reduction rate of UO_3 by hydrogen on the surface area of the starting material has been demonstrated by Notz and Mendel[6] to be linear. The latter authors found that no significant change in surface area took place during the reduction process, so it is perhaps surprising that the relationship shown in Fig. 7 was obtained for this study, which although almost linear does not pass through the origin. It could be that extrapolation of this almost-linear section of this figure to zero surface gives an indication of the nuclei that existed before the reduction process began. The prolonged heating necessary to produce the smaller surface area destroys these high-energy sites, causing the fall in reduction rate with decreasing surface.

Reduction of NiO

The reduction of sample NN_{500} can be seen to be markedly different from the reduction of sample NA_{350}. In particular, for sample NN_{500}, an induction period was observed where no perceptible weight change took place, whereas for sample NA_{350} this phenomenon did not occur. The length of this induction period depended on the experimental temperature and was quite reproducible for a given sample. Vlasenko and Telipko[12] had previously remarked on this induction process, and shown by volumetric measurements that in the initial period of this process rapid adsorption of hydrogen took place due to chemisorption on the nickel oxide surface. The adsorption was then found to slow down while the slower reaction between chemisorbed hydrogen and the nickel oxide surface took place. The process involved would appear to be one of formation of nuclei on the nickel oxide surface. Growth of these nuclei does not begin until a critical concentration is achieved, when three-dimensional growth might be expected to take place.

The sigmoid curves obtained for the reduction of the nickel-oxide samples prepared for this study are quite characteristic of nucleation and growth processes. The data obtained by Delmon[11] was shown to fit a power-law kinetic expression over the acceleratory period of the reaction, but power-law plots of the data shown in Figs. 9 and 12, using powers of t from 2 to 4, were found to be linear over only a small extent of the reaction. The kinetic expression that best fitted the present work was the Avrami—Erofeyev equation, which had been developed to describe reactions where growing nuclei interfered with the growth of neighboring nuclei.

The exponent n in solid-state kinetic expressions is made up of two components, $n = \beta + \lambda$, where β represents the number of stages in nucleation (usually one), and λ represents the dimensions of growth of the nuclei. Growth of nuclei often commences immediately after their formation, so that during the early stages of a reaction for three-dimensional growth n is often $3 + 1 = 4$. In cases where the nucleation process goes to completion before growth of these nuclei commences it might be expected that initially $n = 1$, while later on in the reaction n would equal 3.

Avrami—Erofeyev plots of the data obtained for the reduction of sample NN_{500} are linear up to $\alpha = 0.75$ (Fig. 11) with the exponent $n = 3$, and it would appear that in this case nucleation is complete before growth of these nuclei begins. For sample NA_{350} the different form of the Avrami—Erofeyev equation [Eq. (3)] demonstrates that for the early part of the reaction up to $\alpha \sim 0.25$, $n \sim 1$, the slope of these plots changing to 3 after $\alpha \sim 0.25$; this once more illustrates that growth of nuclei does not begin to any extent until the nucleation process is complete.

The reduction of nickel oxide by hydrogen has been shown by conductivity measurements[17] to proceed by adsorption of hydrogen, as the H_2^+ species, onto cationic sites on the nickel-oxide surface. Reaction between this chemisorbed H_2^+ and neighboring O^{2-} lattice ions follows, producing hydroxyl ions and releasing electrons. These electrons reduce the cationic acceptor sites Ni^{3+}, Ni^{2+}, etc., in the immediate vicinity. Nucleation of the hydroxyl ions so formed and desorption of product water molecules follow. It was also considered quite probable that migration of the H_2^+ species into the lattice interior, and diffusion of OH^- ions by proton exchange between neighboring anions, also took place to a large extent, both processes leading to regeneration of the original Ni^{3+} adsorption site.

It was considered probable in the light of the previously discussed work, and of that of Vlasenko and Telipko,[12] that for sample NN_{500} the diffusion of OH^- ions into the crystal interior takes place to a considerable extent. Anderson and Klemperer[17] found that the reduction process does not proceed until a critical concentration of hydroxyl ions is reached. Presumably these ions have to be in reasonably close contact before their nucleation to product water can take place. The disperse nature of the OH^- ions in the NN_{500} sample while it is being reduced would allow a quite high concentration of this species to build up before formation and desorption of water occurred. This would explain the induction period observed for this sample, where no detectable weight change took place. At the higher experimental temperatures the diffusion of OH^- ions does not have a chance to proceed to the same extent before a critical concentration is formed near the surface.

It has been shown[14] that the residues from nickel-acetate decomposition can contain a considerable excess of oxygen in defects in the nickel-oxide lattice. It is probable that a high proportion of these defects are at or near the crystal surface. Sample NA_{350} has been shown to contain $\sim 2\%$ excess oxygen, and it is thought that on reduction the many surface defect sites will lead to rapid formation of hydroxyl ions in close proximity to each other, leading to the formation of water molecules and their subsequent desorption from the surface. This is observable experimentally as a continuous weight loss from the onset of the reduction process (i.e., no induction period).

This latter process could also result in a high concentration of nickel atoms in the vicinity of the crystal surface and a rapid depletion of the surface cationic adsorption sites, which is thought to explain the higher activation energy observed for this sample.

REFERENCES

1. Cotton and Wilkinson, Advanced Inorganic Chemistry, Interscience, New York (1962), p. 910.
2. J. A. Arfredson, Kgl. Svensk. Vetaskaps. Akad. Handl., p. 413.
3. Harrington and Ruehle, Uranium Production Technology, Van Nostrand (1959).
4. R. M. Dell and V. J. Wheeler, Trans. Faraday Soc., 58, 1590 (1962).
5. R. E. DeMarco and M. G. Mendel, J. Phys. Chem., 64, 132 (1960).
6. K. J. Notz and M. G. Mendel, J. Inorg. Nucl. Chem., 14, 55 (1960).
7. J. W. Mellor, Comprehensive Treatise on Inorganic Chemistry, Vol. XV, Longmans, London (1946).
8. G. Gallo, Ann. Chim. Applic., 17, 535 (1927).
9. G. Parravano, J. Am. Chem. Soc., 74, 1194 (1952).
10. Y. Iida and K. Shimada, Bull. Chem. Soc. Japan, 33, 790 (1960).
11. B. Delmon, Bull. Soc. Chim. France, 1961, 590-597.
12. V. A. Telipko and V. M. Vlasenko, Kataliz i Katalizatory Akad. Nauk UkrSSR Resp. Mezhred Sb., (1965), pp. 146-154.
13. P. S. Clough, D. Dollimore, and P. Crundy, J. Inorg. Nucl. Chem., 31, 361 (1969).
14. J. Leicester and M. J. Redman, J. Appl. Chem., 12, 357 (1962).
15. B. T. M. Willis, Nature, 197, 755 (1963).
16. J. D. Cotton and P. J. Fensham, Trans. Faraday Soc., 59, 1444 (1963).
17. J. S. Anderson and D. F. Klemperer, Proc. Roy. Soc., A258, 350 (1960).

A Temperature Error in the Gravimetric Determinations of Adsorption Isotherms

P. A. Cutting

Gas Council
London Research Station
Michael Road
Fulham, S.W.6, England

ABSTRACT

Adsorption isotherms of a nonporous hydroxylated silica at liquid-nitrogen temperatures are compared. Differences between the isotherms are shown to be due to a difference in temperature between liquid nitrogen and the sample, in the gravimetric method. The temperature difference and the method by which it was measured are reported. It is shown that the temperature difference can be eliminated by use of a special hangdown tube, and a small modification to the Cahn RG Electrobalance used. It is shown that temperature differences between the sample and the environment can be avoided by the use of vapors at room temperature.

INTRODUCTION

It is well known that microgravimetry by means of an electromagnetic balance with automatic recording is a very rapid and versatile method for adsorption studies, but volumetric methods are widely used also. In this paper a comparison is made between isotherms obtained by the use of both methods at liquid-nitrogen temperature with the same nonporous material. On comparison it is found that the isotherms were not superimposable, the gravimetric isotherm showing lower adsorption.

EXPERIMENTAL

The material studied was an hydroxylated silica, TK800, which has been shown by different methods to be nonporous; the confirmation of the results was made by using another nonporous hydroxylated silica, Fransil.[1] The volumetric isotherm was determined by use of a Sing — Harris-type apparatus[2] and by other co-workers using a Lippens — Linsen — DeBoer-type apparatus.[3] The gravimetric studies were mostly carried out on a Cahn RG Electrobalance with nichrome hangdown wires and gold buckets, the hangdown tube being immersed in a liquid nitrogen bath. The isotherms at 15- and 20-cm depth of immersion in the bath were confirmed by using a Sartorius Electronobalance. The nitrogen used for adsorption in both the volumetric and gravimetric work was pure and dry; that for the bath was commercial liquid nitrogen. Pressure measurements were made to better than ±0.1 torr with a C.E.C. pressure transducer, 0-15 psia, type 4/326/L210, which had been calibrated against a mercury manometer.

MODIFICATION OF GRAVIMETRIC APPARATUS AND RESULTS

In Fig. 1 the volumetric and gravimetric isotherms at depth of immersion of 20 cm are shown. Since the gravimetric isotherm cuts the ordinate at $p/p_0 = 1$ and the volumetric isotherm approaches it asymptotically, the difference between the isotherms was thought to be due to a difference in temperature of the two samples. The temperature of the gravimetric sample was measured with chromel — alumel thermocouples, one centrally inside the 25-mm diameter tube and one outside, both 20 cm below the level of the top of the liquid in the nitrogen bath. The potential difference between the thermocouples was found to be 12 μV, corresponding to the temperature inside the tube being 0.7 C higher than that outside.[4] This difference was found to be the same whether the tube was evacuated or filled with nitrogen to its saturation point p_0, as shown by condensation in the bottom of the tube.

From tables it was calculated that 0.7 C rise in temperature gives an increase of about 100 torr in the saturation vapor pressure of nitrogen.[5] When this correction to the saturation vapor

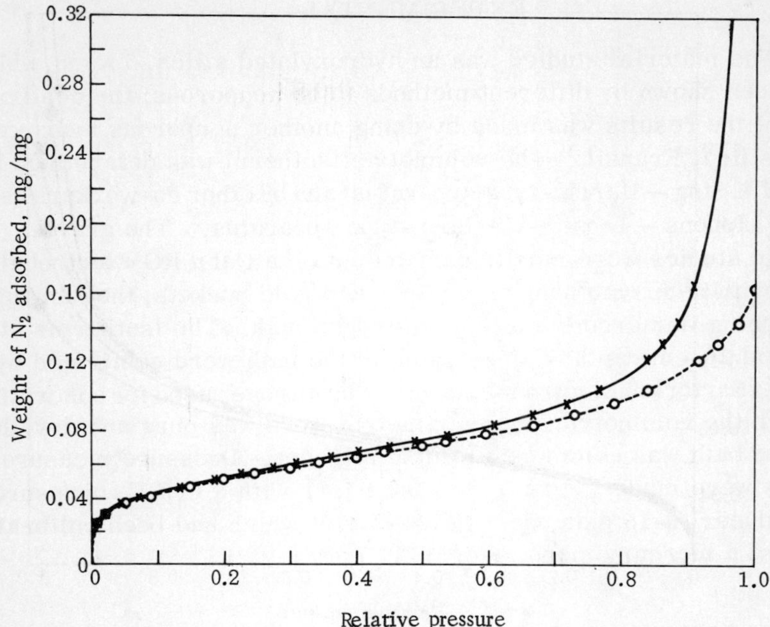

Fig. 1. Comparison of volumetric and gravimetric nitrogen isotherms. Dotted line, gravimetric, 20 cm depth; solid line, volumetric; ×, points corresponding to ○ after correcting for temperature error.

pressure was applied to the pressure readings, the resulting isotherm became coincident with the volumetric one, as shown in Fig. 1 by the points marked × on the volumetric isotherm. The upper limit of measurement of relative pressure under the above conditions is thus 0.9. If the bucket was at a depth of 15 cm, the temperature error was calculated to be greater than 1 C by using the equation[6]

$$\log p_0 = (-339.8/T) + 7.71057 - 0.0056286T$$

Increasing the depth of liquid nitrogen to 30 cm gave some improvement, as shown in Fig. 2. Hausmann[7] suggested that equilibria could be achieved more rapidly by bringing the sample bucket into better thermal contact with the outer liquid-nitrogen bath. He did

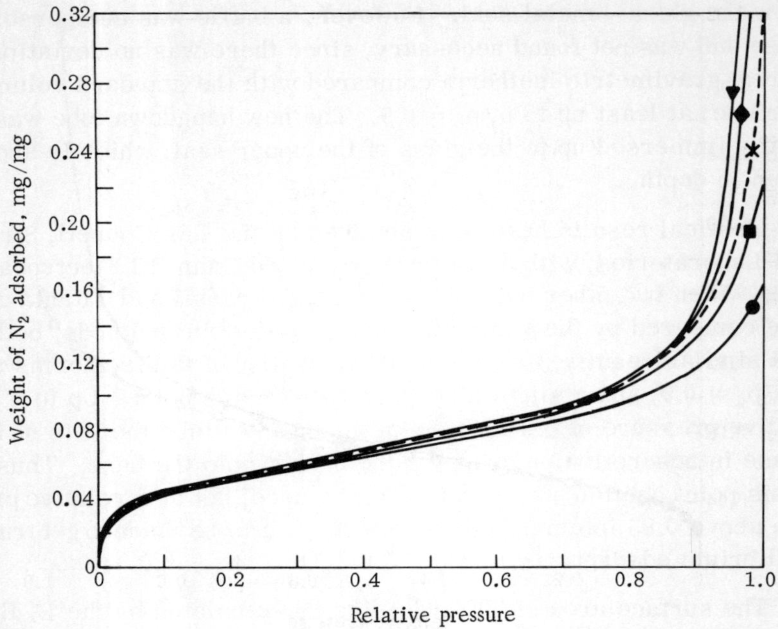

Fig. 2. Comparison of nitrogen isotherms: ▼) volumetric; ◆ new hangdown tube; × flanged tube; ■ 25-cm-diameter tube; 30 cm depth; ● 25-cm-diameter tube, 20 cm depth.

this by having a metal plate on a spring on which the bucket rested during adsorption only, the other end of the spring being attached to a tungsten rod passing through the glass. This principle was modified and used to make a hangdown tube with a solid metal base and a large heat sink above it in the form of a metal flange. The bucket was suspended some 4 mm above the metal base. The current from the phototube of the Cahn RG Electrobalance to the servomechanism could be interrupted by a simple mechanical switch, so that the sample arm of the balance was lowered to allow the bucket to sit on the base of the hangdown tube during adsorptions. To carry out the weighing, the switch was closed to raise the arm. This procedure gave further improvement, as can be seen in Fig. 2.

Further improvement was obtained by use of the modified hangdown tube, as shown in Fig. 3. The Kovar-metal portion in the middle was made narrow (15 mm in diameter) to minimize convection[8] and a baffle could be inserted at the top of this portion, just below the glass — metal seal. However, a baffle was cumbersome to use and was not found necessary, since there was no deviation of the new gravimetric isotherm compared with the standard volumetric one, at least up to $p/p_0 = 0.9$. The new hangdown tube was usually immersed up to the glass of the upper seal, which is about 30 cm in depth.

Identical results have been obtained by the Gas Council, Stretford Laboratories, with a similarly modified Cahn RG Electrobalance. When two other nonporous samples, Fransil and Titania P25, were compared by the volumetric and gravimetric methods, both gave similar results, i.e., a complete identity of the isotherms up to $p/p_0 = 0.9$, and a slight divergence above this point. Up to a relative pressure of 0.8 with the new hangdown tube there is no increase in adsorption on dropping the bucket onto the base. Thus to this point continuous weighing can be used, but at a relative pressure above 0.85 intermittent weighings have to be made to get true equilibrium adsorptions.

The surface area of TK 800 silica as calculated by the B. E. T. method[9] was $154\,m^2 \cdot g^{-1}$ from volumetric and corrected gravimetric measurements, and $153\,m^2 \cdot g^{-1}$ when measured at 20-cm depth of immersion in the unmodified apparatus. Complete agreement was

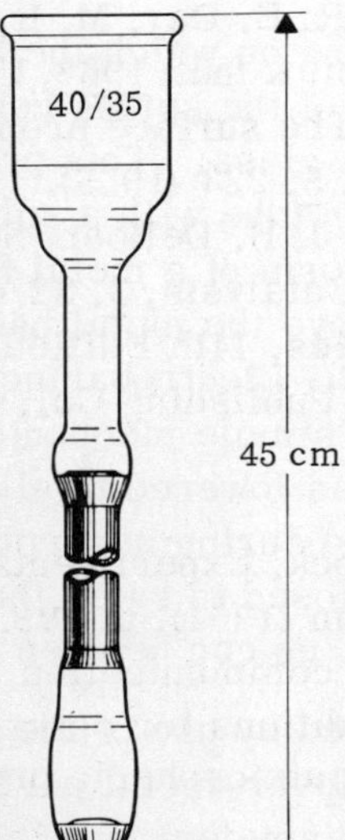

Fig. 3. New hangdown tube.

obtained by the volumetric and gravimetric methods with the new hangdown tube, for the Fransil (38.7 $m^2 \cdot g^{-1}$) and the Titania (43.2 $m^2 \cdot g^{-1}$). Although the lowering of adsorption has little effect on the surface area, it will have a more marked effect on pore-size distribution: it will show an apparent decrease in the distribution of pore size at the higher relative pressures, and may also give an apparent increase in microporosity.[10]

An alternative method of avoiding the temperature error is to work at about room temperature and use a suitable adsorbate. Carbon tetrachloride, which has a saturation vapor pressure of 100 torr at 23 C has been used successfully in this way on all the above samples.[11] With microporous materials, because of enhanced micropore filling at low relative pressures, this technique gives meaningless surface-area values, but it does give reliable micropore volumes.

ACKNOWLEDGMENTS

I would like to thank N. Zivy et Cie. S. A. of Oberwil-Basel for the use of their Automatic B. E. T. Apparatus (after Dr. Hausmann); Cahn Instrument Co., Ltd., for their cooperation throughout; and The Gas Council for their permission to publish these results.

REFERENCES

1. J. D. Carruthers, P. A. Cutting, R. E. Day, M. R. Harris, S. A. Mitchell, and K. S. W. Sing, Chem. & Ind., 1968, 1772.
2. M. R. Harris and K. S. W. Sing, The surface properties of precipitate alumina, J. Appl. Chem., 5, 223 (1955).
3. B. C. Lippens, B. G. Linsen, and J. H. DeBoer, Studies on pore systems in catalysts, 1, J. Catalysis, 3, 32 (1964).
4. Handbook of Chemistry and Physics, 44th Edition (C. D. Hodgman, ed.), The Chemical Rubber Publishing Co., Cleveland, Ohio, U. S. A. (1962), p. 2702.
5. Ibid., p. 2439.
6. H. W. Melville and B. G. Gowenlock, Experimental Methods in Gas Reactions, Macmillan, London (1964), p. 198.
7. T. Hausmann, Lenzburg, Private communication.
8. L. Cahn and N. C. Peterson, Conditions for optimum sensitivity in thermogravimetric analysis at atmospheric pressure, Anal. Chem., 39, 403 (1967).

9. S. Brunauer, P. H. Emmett, and E. Teller, Adsorption of gases in multimolecular layers, J. Am. Chem. Soc., 60, 309 (1938).
10. K. S. W. Sing, Chem. & Ind., 1967, 829.
11. P. A. Cutting and K. S. W. Sing, Chem. & Ind., 1969, 268.

Determination of the Diffusion Coefficient of Vapors by Means of a Microbalance

K. M. Dijkema and J. C. Stouthart

Physics Department
Eindhoven University of Technology
Eindhoven, Netherlands

ABSTRACT

Determination of the diffusion coefficient of vapors in gases by Stefan's method leads to results of rather low reliability. Improvement can be attained by using a recording microbalance. The determination is performed by recording the loss of weight of a small bucket containing a liquid evaporating through a narrow vertical tube. The suspension wire is placed in the axis of the tube. In the case of very narrow tubes ($\varnothing < 10^{-3}$ m), tube and container are weighed as a whole.

Variations in apparent weight may occur due to variations in the room temperature and to electrical charges. Small variations in temperature of the evaporating liquid have a very strong influence, especially at higher evaporation rates.

The diffusion coefficient of water vapor in air is determined at 323, 333, and 343 K.

INTRODUCTION

The coefficient of diffusion for vapors in gases can be determined by the method of Stefan. A certain amount of liquid is put in a vertical, transparent tube. A stream of gas at the upper end of the tube causes an evaporation of the liquid at a certain rate, which can be measured by determining the height of the liquid column at regular intervals.

The diffusion coefficient can be calculated from the following equation:

$$D = \frac{Gl RT}{MtAp \ln[p/(p - p_s)]} \tag{1}$$

where G is the mass of evaporated liquid, l is the length of the gas column, A is the cross-sectional area of the tube, and p_s is the saturation vapor pressure of the liquid at T (K), while the other symbols have their usual meaning.

The determination of D in this way leads to results which show a poor degree of accuracy. Deviations from the true value might be due to the relatively low degree of accuracy that can be obtained in determining small differences in height. It was expected that a determination by means of weighing the residual water at regular times would give more accurate results. This proved not to be the case. Subsequently we used a recording electrical microgram balance for continuous measuring of the evaporation rate. This method is described in the paper.

DETERMINATIONS WITH AN ANALYTICAL BALANCE

Description

At the bottom of a vertical tube of stainless steel (Fig. 1) with a length of 40 cm and a diameter of 1 cm a small glass bucket was introduced by means of a thin ($\varnothing = 100\ \mu$) platinum suspension wire. The bucket was nearly completely filled with distilled water. The tube was surrounded by a mantle through which water of constant temperature was flowing.

At the upper end of the tube was situated a hollow ring with small bores at the inner side. The arrangement was such that from each of these bores a small stream of air could be led over the upper end of the tube. The air streams were blowing upwards at an elevation angle of 45°.

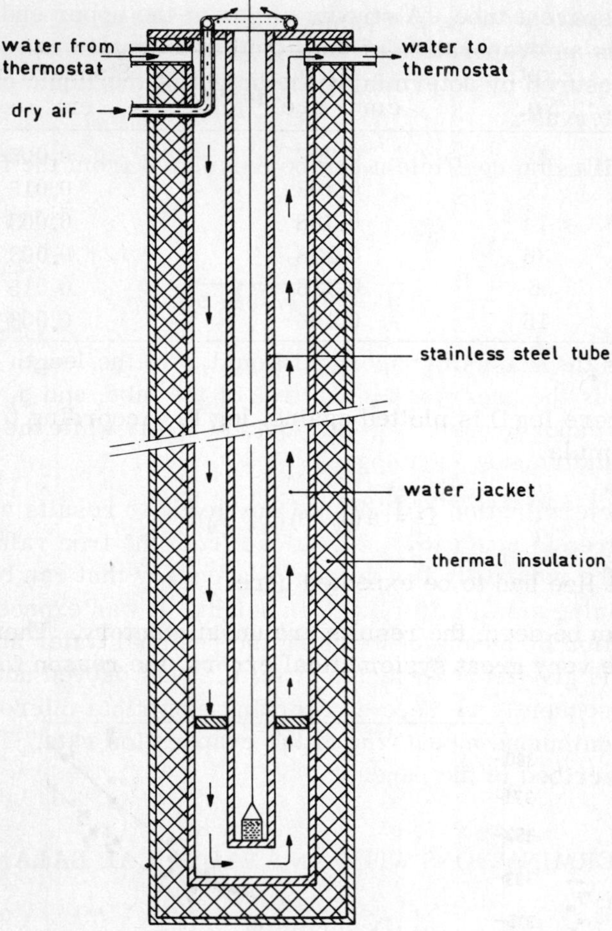

Fig. 1. Diffusion column used in the preliminary determinations.

The bucket was weighed before it was placed in the tube. After the evaporation had proceeded for 4 to 50 h, the bucket was taken out of the tube and covered with a small weighed glass plate. Corrections for errors caused by manipulating the bucket were made. These were determined by experiments of short duration.

Results

Measurements were made at 315, 323, 328, 337, 343, and 357 K. The results are shown in Table I. These results are also shown in

Table I

Temp., K	Expt. No.	D, $cm^2 \cdot sec^{-1}$	Standard deviation, $cm^2 \cdot sec^{-1}$
315	12	0.297	0.005
323	5	0.308	0.015
328	14	0.328	0.003
336.6	36	0.328	0.003
343	6	0.336	0.015
357	15	0.406	0.003

Fig. 2, where log D is plotted against log T. According to a well-known formula,

$$D = a(p_0/p)(T/T_0)^b \tag{2}$$

A straight line had to be expected here.

As can be seen, the results are unsatisfactory. There have been some very great systematical errors, the reason for which

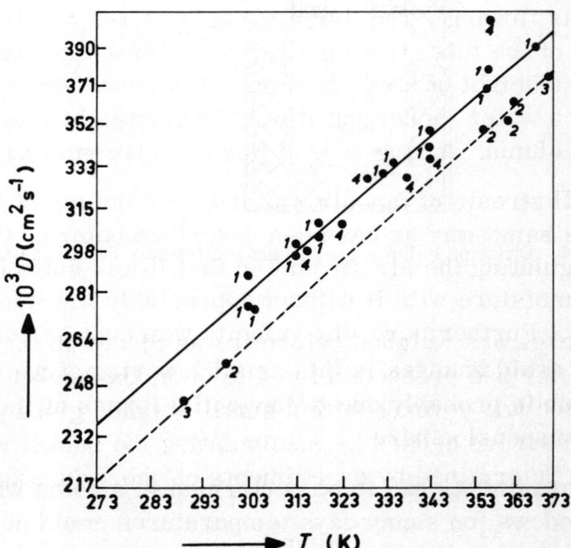

Fig. 2. Diffusion coefficient versus T; logarithmic scale. Experimental values of D: 1) Schirmer (solid line); 2) Mache and 3) Trautz and Müller (dashed line); 4) Dijkema and Stouthart (I).

we have not been able to trace. Compared to the results of others,[1,6,7] the measurements at 357° gave too-high values. The other measurements compare rather well with the experiments of Schirmer.[1] All these measurements lie reasonably well on a straight line, but are somewhat higher than most values found by other authors.

MEASUREMENT WITH A RECORDING MICROGRAM BALANCE

Experimental Setup and Procedure

A small bucket (height 10 mm, diameter 8 mm) was suspended from one arm of a recording microgram balance by means of a thin (25 μ) platinum wire. This wire was placed in the axis of a tube made of stainless steel. The tube was 30 cm long and had an inner diameter of 3 mm. It was carefully adjusted in a perpendicular position. The balance was placed on two crossed micrometer carriages, which allowed carefully controlled movements in a horizontal plane. This enabled the wire to be centered in the tube. To avoid undesirable contact of the wire and the inner wall of the tube the deviations from the horizontal position of the beam of the balance were kept very small. The vertical tube was surrounded by a water jacket through which water of constant temperature ($\pm$0.01-0.002 K) was flowing. The top of bucket was 1-2 mm beneath the bottom end of the tube, in a small space with a diameter of about 12 mm and a height of about 15 mm. This space again was surrounded by a water jacket and it was pressed tightly to the lower end of the column. A ring of lead was used for sealing (Fig. 3).

A small stream of dry air was led over the top of the column in about the same way as has been described before. Care was taken in regulating the air stream so that it was sufficiently strong to take all moisture with it without appreciably lowering the apparent weight. Furthermore, the velocity was kept as constant as possible to avoid changes in this small lowering of weight (about 10 μg), which is probably due to tangential forces of the stream of air on the suspension wire.

The experiments had to be performed in a room where, owing to large windows, on sunny days temperatures could not be kept under control otherwise than by means of an irregular stream of cold air, which caused variations in the balance equilibrium. To diminish these uncontrollable changes in weight a hood of methylmetacrylate (plexiglass) was placed over the balance. Electrosta-

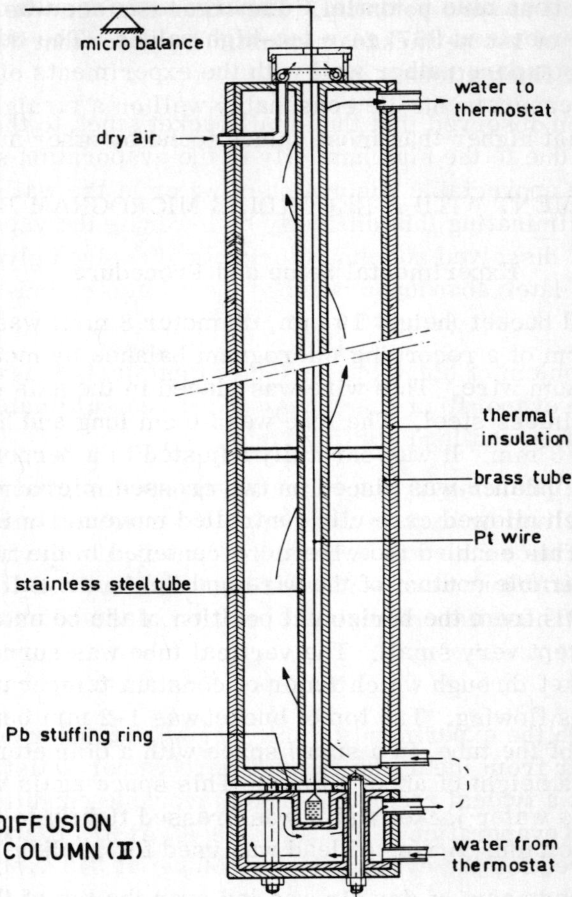

Fig. 3. Diffusion column for continuous determination of the evaporation by means of a microbalance.

tic charges then caused irregularities, which we tried to eliminate by a radioactive capsule containing 100 μg of ^{226}Ra. This indeed had a very strong influence, even to a greater extent than we had expected. It may be due to ionization of the ambient air. The balance itself was at a potential of 6 V against earth potential. The ionization may have caused a small leakage current, simulating a certain change of weight.

Finally, we obtained satisfactory results by withdrawing the radioactive source and covering the inside of the plastic hood with aluminum foil at zero potential. The outer surface was covered with a layer of 1-cm-thick foam-plastic insulator that diminished temperature fluctuations.

We often observed that the small bucket stuck to the wall. This is probably due to the high humidity in the evaporation space, which might cause appreciable adsorption of water at the walls. We succeeded in eliminating this difficulty by lowering the vapor pressure by means of dissolved sodium chloride or dissolved glycerol. This method was later abandoned because the concentrations of the solution varied.

The problem of obtaining perfectly tight seals between the column and the evaporation space has not yet been ultimately solved. We used Teflon and lead stuffing rings.

In some preliminary experiments the stream of air was replaced by helium or carbon dioxide. After several hours a constant evaporation rate was achieved. The calculated values of the diffusion coefficient were somewhat higher than those found in the literature. These experiments will be repeated and extended.

Results

Although the experiments were carried out under conditions that were far from ideal, the results were rather satisfactory. Figure 4 shows a typical recording under favorable conditions for the diffusion and evaporation at 313 and 343 K. At 313 K about 45 min were required for the evaporation of 100 μg; at 343 K, about 10 min were required.

This factor of 4.5 is in agreement with Eq. (1); i.e., for $p_s \ll p$, the pressure term $p \ln p/(p - p_s)$ in the denominator of (1) can be approximated by p_s. The ratio of the saturation vapor pressures at 343 and 313 K is 4.2.

Such recordings show clearly whether or not the evaporation rate has been constant during the experiment. It proved that this was the case 1-2 days after starting the experiment. The long period of nonstationary evaporation may be partly due to the thermal resistance between the walls of the evaporation room and the suspended container.

The small irregularities at 313 K and the greater disturbances at 343 K must be due to changes in the temperature of the water

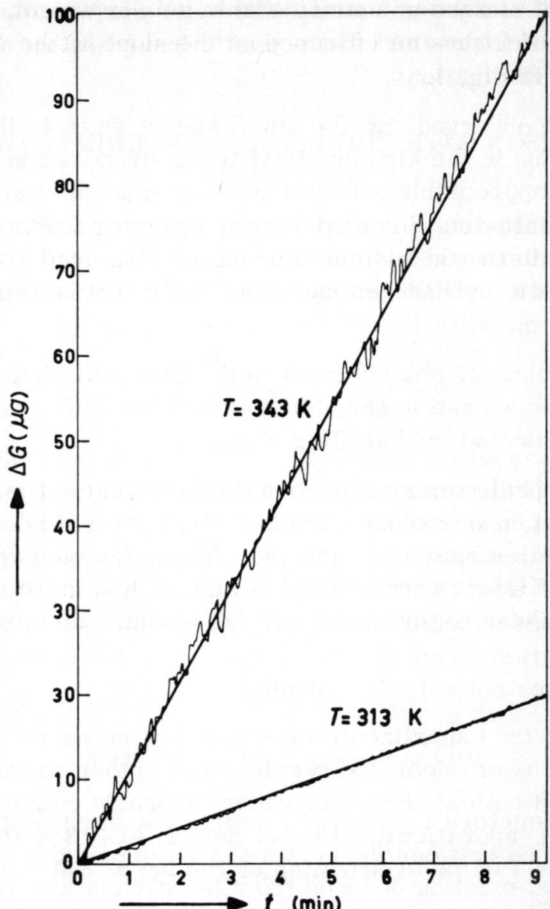

Fig. 4. Typical recording of the evaporation at 313 and 343 K.

mantle. This temperature did not change more than ±0.01 K, but apart from a small lag in time we found good correlation between this temperature change and the apparent changes in weight as they are found in the recording of the evaporation line.

It seems astonishing that such small variations in temperature should cause the evaporation of such great quantities of water. We feel that the variations described must be due to a "jet" effect of the evaporating water molecules, escaping at a velocity of about 600 m/sec and giving a momentum to the balance in the opposite direction that simulates a considerable gain in weight. However, this takeoff effect has no influence on the slope of the weight versus time curves in Fig. 4.

COMPARISON WITH THEORY AND EXPERIMENTAL DATA OF OTHERS

Kinetic theories of transport phenomena in dilute gases, as can be found in the treatise *Molecular Theory of Gases and Liquids* by Hirschfelder et al.,[2] lead to the following expression for the diffusion coefficient of two gases:

$$D = \frac{3kT}{16n} \frac{m_1 + m_2}{m_1 m_2} \frac{1}{\Omega_{12}} \tag{3}$$

Here Ω_{12} is what is termed a collision integral. The value of this collision integral depends on the force field between the colliding molecules, and may be calculated from the collision integrals of the pure components. In this calculation the assumption must be made that the field of force between a polar and a nonpolar molecule can be described by the same expression as can be used for the forces between nonpolar molecules, i.e., the well-known Lennard-Jones potential equation:

$$\rho(r) = 4\varepsilon_{12}\{(\sigma_{12}/r)^{12} - (\sigma_{12}/r)^6\} \tag{4}$$

The parameters ε_{12} for the minimum of the potential and σ_{12} for the distance between the centers of the molecules at this minimum potential must be deduced from the parameters for the pure components by

$$\varepsilon_{12} = (\varepsilon_1 \varepsilon_2)^{\frac{1}{2}} \tag{5}$$

and

$$\sigma_{12} = \tfrac{1}{2}(\sigma_1 + \sigma_2). \tag{6}$$

The parameters for the pure components must be calculated from experimental data. The parameters for air may easily be found in literature. For water various values may be found. The lowest (and probably the best) values are those given by Rowlinson and Townley[3]:

Table II

Temp., K	Number of recordings	D, $cm^2 \cdot sec^{-1}$	Standard deviation
323	26	0.290	0.003
333	10	0.309	0.001
343	19	0.325	0.002

$$\varepsilon_1/k = 356 \text{ K and } \sigma_1 = 2.649 \times 10^{-10} \text{ m for water vapor}$$

$$\varepsilon_2/k = 97 \text{ K and } \sigma_2 = 3.617 \times 10^{-10} \text{ m for air}$$

Measurements were made at 323, 333, and 343 K. The results are shown in Table II and in Fig. 5.

De Vries and Kruger[4] used these data to calculate the diffusion coefficient for water – air between 273 and 373 K. They found the expression

$$D = 0.212\,(p_0/p)\,(T/T_0)^{1.88} \tag{7}$$

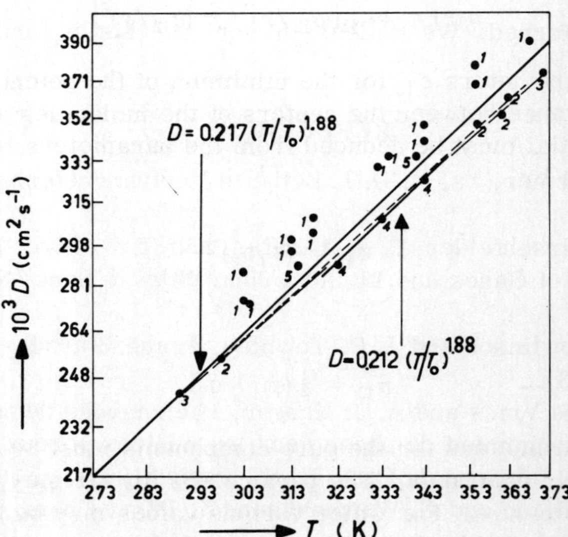

Fig. 5. Experimental and theoretical values of D. 1) Schirmer; 2) Mache (dashed line); 3) Trautz and Müller (broken line); 4) Dijkema and Stouthart (II) (solid line); 5) Le Blanc and Wupperman.

but on the basis of an estimated uncertainty of about 4% in this value, and on the basis of a most-probable value of D = 0.257 cm^2/sec at 298 K (theoretical value 0.251 $cm^2 \cdot sec^{-1}$) deduced from measurements of 13 different authors, they proposed

$$D = 0.217(p_0/p)(T/T_0)^{1.88} \tag{8}$$

However, there seems no reason for this correction. When those measurements are taken into account which show a dependence on temperature that differs no more than about 4% from the calculated value, as well as those measurements that differ less than about 4% from the most-probable value at 298 K, only the measurements of Le Blanc and Wuppermann,[5] of Mache,[6] and of Trautz and Müller[7] need to be considered (Fig. 5). The measurements of Le Blanc and Wuppermann are somewhat higher than the corrected theoretical value of De Vries and Kruger.[4] Those of Mache and of Trautz and Müller are in better agreement with the uncorrected value, as are our measurements, which are also shown in Fig. 5.

From these considerations, and from the experience that positive deviations are more likely to occur than negative ones, we deduce that the uncorrected value

$$D = 0.212(p_0/p)(T/T_0)^{1.88} \tag{9}$$

is to be preferred. We shall try to confirm this by further experiments.

REFERENCES

1. R. Schirmer, Zs. V. D. I. Beihefte Verfahrenstechn., 6, 170 (1938).
2. J. O. Hirschfelder, C. F. Curtiss, and R. B. Bird, Molecular Theory of Gases and Liquids, John Wiley & Sons, New York (1963).
3. J. S. Rowlinson and J. R. Townley, Trans. Faraday Soc., 49, 20 (1953).
4. D. A. De Vries and A. J. Kruger, Phénomènes de transport avec changement de phase dans les milieux poreux ou colloidaux, Editions du C. N. R. S., Paris (1967), p. 160.
5. M. Le Blanc and G. Wuppermann, Z. Phys. Chem., 91, 143 (1916).
6. H. Mache, Math. Naturw. Sitz. Ber. Wien, 119, 1399 (1910).
7. M. Trautz and W. Müller, Ann. Phys., 414, 332 (1935).

Activation of Cellulose-Triacetate Carbon by Reaction with Carbon Dioxide: A Microgravimetric Study

N. G. Dovaston
Sunderland Polytechnic
Sunderland, England
and
B. McEnaney
Bath University of Technology
Bath, England

ABSTRACT

An apparatus incorporating a Cahn RG electrobalance is described which follows the reaction of a cellulose-triacetate carbon with carbon dioxide at 76 torr and temperatures up to 1300 K, and measures the porosity so developed in the carbon by adsorption of carbon dioxide at 195 K. Reaction rates obtained for the activation process yield kinetic parameters which indicate that below about 1208 K the reaction rate is chemically controlled, the influence of mass transport of gases being negligible. A measure of the microporosity developed on activation is obtained by plotting the 195 K adsorption isotherms according to the Dubinin Equation. The influence of heat treatment of the carbon on the activation kinetics and porosity development is considered.

INTRODUCTION

With the continuing technological importance of activated carbons and with possibilities of their wider application (e.g., by development of molecular-sieve materials from polymer carbons[1]) there is increasing need to understand all factors which determine their adsorptive properties. Much work has been published on the kinetics of reaction of oxidizing gases with carbons and graphites, particularly with reference to fuel and nuclear technology. However, relatively little work has been published which attempts to relate the development of porosity in activated carbons to the kinetics of the activation process. Yet such information is necessary if efficient use of resources in activated-carbon manufacture is to be made. The work reported in this paper is part of such an investigation.

Previous work[2] has established that development of porosity in carbons by reaction with carbon dioxide depends markedly on the graphitic character of the carbon. Nongraphitizing carbons were shown to be more reactive and to develop greater porosity on activation than graphitizing carbons. This paper reports preliminary results of an investigation of the influence of heat treatment of a nongraphitizing carbon to 2300 K on the development of porosity by reaction with carbon dioxide, using a microgravimetric technique. Attention is confined in this paper to carbons heat-treated to 1473 K, with emphasis on experimental technique. It has been shown[3] that for unactivated nongraphitizing carbons, the predominant process on heating from 1200 to 3000 K is a reduction in open-pore volume. This appears to be a combination of elimination of pores by sintering and closure of open pores to produce appreciable closed-pore volume in the high-temperature carbons, but the relative contribution of these two factors has not been investigated. The technique of redevelopment of porosity by activation may throw more light on the mechanism of this process.

CELLULOSE-TRIACETATE CARBON

Cellulose-triacetate carbon was chosen for study for a number of reasons. In contrast to cellulose, fusion occurs during the carbonization of cellulose triacetate to produce a coke which unlike many cokes is nongraphitizing.[4] Kipling and Shooter[5] have suggested that formation of a graphitizing carbon requires production of polycyclic aromatic molecules at an early stage of the carbonization and fusion between 670 and 820 K, which permits orientation of these molecules to form the nuclei of graphitic microcrystals. It has been shown[6] that formation of these nuclei starts by separation of anisotropic droplets from an isotropic matrix during the fluid stage of carbonization; these droplets contain the polynuclear aromatic molecules in lamellar array. In the case of cellulose triacetate the fluid stage occurs below 600 K, which is too low for the graphitizing process to be initiated.

Bailey and Everett[7] have shown that the adsorptive properties of a polyvinylidene-chloride carbon are related both to the morphology of the original polymer and the conditions during carbonization. Evidence from optical and scanning electron microscopy[8] suggests that, as a result of fusion during carbonization, cellulose-triacetate carbon has a glasslike structure, and that therefore the properties of the carbon are independent of the morphology of the polymer. Cellulose-triacetate carbon also develops a relatively high open and closed porosity on heat-treatment to 2000 K,[3] which also commends it as a material for study.

CARBON PREPARATION

It is clear that it is important to closely specify the thermal history of carbons in order to ensure reproducible materials, and therefore all stages of carbon production have been carefully controlled. Cellulose triacetate was synthesized by acetylation of cellulose powder (Chromedia CF 11, Whatman Co. Ltd.)[9]; yields were 90% of the theoretical amount and assays indicated triacetylation. A constant starting weight (4 g) of polymer was heated in a metered stream of oxygen-free nitrogen at a linear rate (3.5 $deg \cdot min^{-1}$) to 1229 or 1473 K, held at that temperature for 30 minutes, and then cooled; the apparatus has been described elsewhere.[10] A reproducible yield of carbon (15 ± 0.5%) was obtained; the carbon was ground to between 20 and 40 BSS mesh.

THE MICROGRAVIMETRIC APPARATUS

The apparatus was designed to follow gravimetrically, by means of a Cahn RG electrobalance, both the activation of the carbon by reaction with carbon dioxide at about 1200 K and the adsorptive capacity of the carbon at 195 K. Activation and adsorption measurements could be carried out consecutively in order to avoid pickup of surface oxides by the carbon on exposure to atmosphere. The apparatus could readily be adapted for general use in the study either of the influence of heat treatment or of high-temperature reaction with gases on the surface properties of materials which are reactive in atmospheric conditions.

The apparatus is mounted on a heavy welded-steel frame bolted to the wall and floor of the laboratory. The electrobalance is attached to this frame by a metal clamp which is also bolted to the wall. The orientation of the beam of the balance is perpendicular to the wall, since this arrangement was found to improve substantially the stability of the instrument compared with a parallel orientation to the wall.

When the apparatus is used for activation measurements (Fig. 1) carbon dioxide (99% pure) supplied from cylinders is purified by passage over an activated molecular sieve A (Linde, Type 3A), and then passes through a precision needle valve B (Genevac Ltd.) into a section maintained at atmospheric pressure; any excess gas escapes through a mercury blowoff (not shown). The flow rate is measured on an interchangeable jet flowmeter C and controlled by needle valve D, which admits the gas into the reaction system. The activating gas is drawn over the sample by pump M, pressure in the reaction vessel (76 torr) being controlled by needle valve L, and measured on manometer F.

The reaction vessel and thermocouple sheath are made from vitreous silica, the sheath being attached to the vessel by a B19

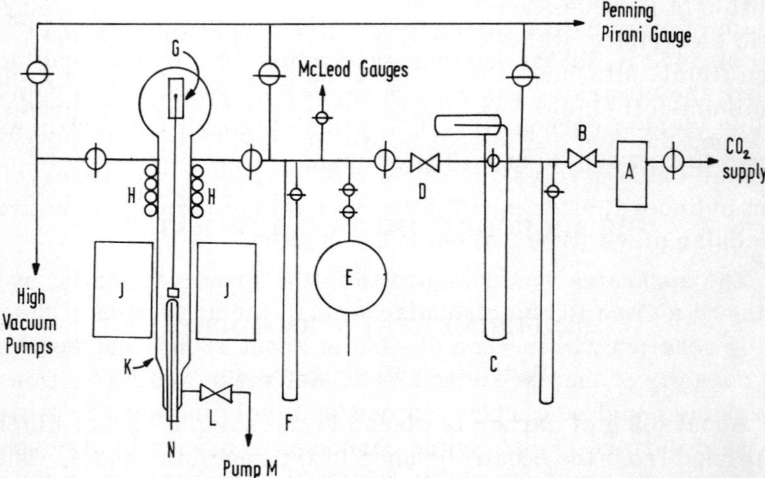

Fig. 1. Schematic diagram of the apparatus. A) molecular sieve; B, D, L) needle valves; C) flowmeter; E) reservoir; F) manometer; G) electrobalance; H) cooling coils; J) demountable furnace; K) silica reaction vessel; N) thermocouple.

cone-and-socket joint. A Pt/Pt-13%Rh thermocouple measures the sample temperature during activation. The hangdown wire and sample bucket were fabricated from platinum. The demountable furnace was constructed in two halves hinged together and utilizes two flexible heating elements (Electrothermal Ltd.). To ensure that all carbons are subject to the same thermal history, the furnace is heated at a linear rate (10 deg · min^{-1}) to the activation temperature by means of a potentiometric program controller (Transitrol Type 1294) installed in the primary circuit of the transformer which supplies power to the furnace elements. During the operation of the furnace the upper part of the reaction vessel is cooled by water circulation through a copper spiral H near and around the B40 cone-and-socket joint that connects the reaction vessel to the electrobalance.

For adsorption measurements the furnace is replaced by a low-temperature bath constructed from a polythene container in two halves which bolt together to fit closely around the reaction vessel. When suitably lagged and filled with solid carbon dioxide the bath gives a temperature of 195 K; the Pt/Pt-13%Rh thermocouple is replaced by a chromel — alumel thermocouple for low-temperature measurement. It was found necessary to periodically repack the bath with solid carbon dioxide to maintain constant temperature. Addition of solid carbon dioxide to the bath was found to generate static electricity, which caused the platinum sample bucket to become firmly attached to the wall of the reaction vessel; a grounded aluminum-foil shield was interposed between the solid carbon dioxide and the reaction vessel to prevent this from occurring. The carbon dioxide used as adsorbate was supplied to the reservoir E from cylinders, after purification by passage through an activated molecular sieve (type 3A) cooled by a salt — ice mixture.

EXPERIMENTAL PROCEDURE

Activation Measurements

About 0.5 g of carbon is placed in the sample bucket which is suspended from the electrobalance in the reaction vessel. The system is evacuated and the sample heated at a linear rate to 573 K, with pumping when outgassing commences. Outgassing is continued until a vacuum of 10^{-6} torr is obtained. The sample is then heated at a linear rate to the activation temperature, when further weight loss occurs due to loss of surface oxides. When a steady weight is recorded the system is isolated from the high-vacuum pumps and a flow of 200 ml · min^{-1} (STP) of carbon dioxide is introduced until the pressure in the reaction vessel reaches 76 torr; this pressure is maintained constant by adjusting needle valve L. Adjustment of the pressure takes less than 30 sec. Weight loss during activation is followed continuously on the electrobalance recorder. After the required amount of carbon has been burned off, the activation is discontinued by isolating the gas supply and pump M and evacuating the reaction vessel with the high-vacuum pumps.

Adsorption Measurements

When the reaction vessel has cooled, the furnace is removed and the low temperature bath fitted. Carbon dioxide is admitted to the reaction vessel from the reservoir, the initial aliquot being chosen to give an equilibrium pressure of about 0.01 torr for the first point on the adsorption isotherm. Weight increases due to adsorption are measured on the electrobalance, the recorder trace giving a good indication of the equilibrium condition. Equilibrium pressures are measured on two McLeod gauges covering the range 10^{-6} to 10 torr, higher pressures being measured on the mercury manometer F. Measurements of adsorption are continued to about 600 torr; desorption isotherms may then be determined.

It was considered most practicable to operate the balance at a sensitivity of $\pm 100\ \mu g$ for activation measurements, and $\pm 20\ \mu g$ for adsorption measurements, although greater accuracy was possible for both operations ($\pm 1\ \mu g$ for adsorption measurements). During activation, gas flow is downward over the sample, which experiences both the piston effect of gas flow and the buoyancy effect of the denser medium. The combined effect produces an initial weight gain of about 2 mg. Buoyancy effects during adsorption measurements were found to be normally less than 1% of the amount adsorbed at the highest pressures used. Errors in weight measurements during adsorption due to thermal transpiration are barely significant, and in any case only relevant to the initial part of the isotherm in the range 0.01 to 0.1 torr.[11] Corrections to pressure measurements due to thermal transpiration have been calculated using the data for carbon dioxide of Bennet and Tompkins.[12] Under the conditions employed the correction becomes less than 1% for pressures greater than 0.02 torr.

RESULTS AND DISCUSSION

Activation Measurements

The cellulose-triacetate carbons which were heat treated to 1229 and 1473 K were activated a various temperatures between 1237 and 1164 K. Linear weight-loss curves (Fig. 2) were obtained at all temperatures for both heat-treated carbons over the whole range of burnout examined (up to 40%), and were thus of the form w = k't, where w is the weight loss at time t, and k' is the slope of the weight-loss curve. The reaction rate k is expressed in terms of the original weight of carbon (units: g/sec/g of unactivated carbon). Experimental kinetic parameters were determined by a single-run method. The carbon was initially activated at about 1235 K until sufficient data were obtained to determine the reaction rate. The temperature was then reduced in steps of about 10 deg to about 1160 K, and the reaction rates determined at each temperature; a higher temperature was occasionally reimposed as a check on reproducibility.

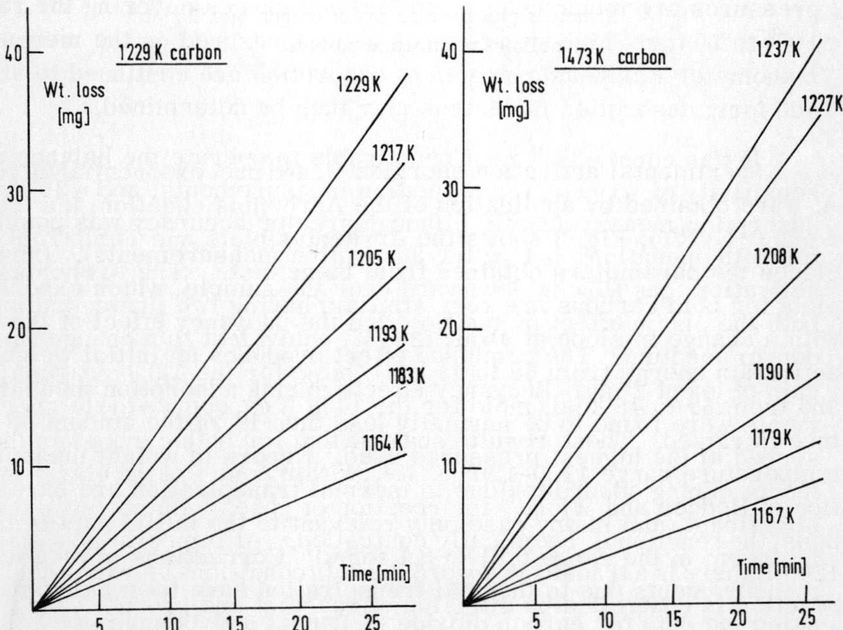

Fig. 2. Weight-loss curves for reaction of carbon dioxide with cellulose-triacetate carbons.

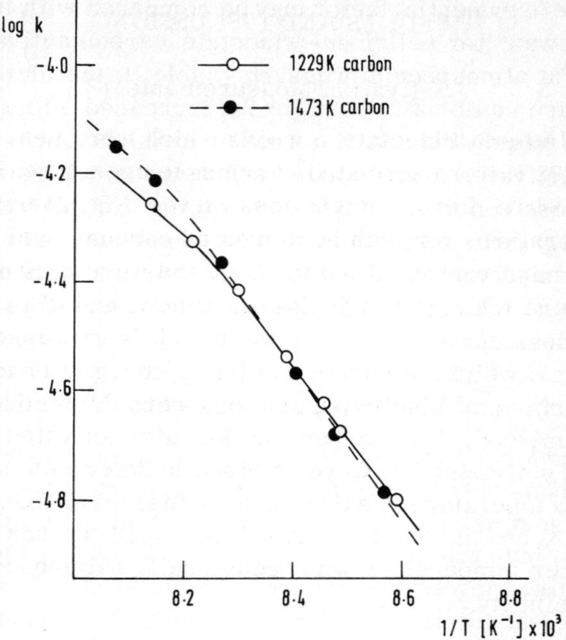

Fig. 3. Arrhenius plot for reaction of carbon dioxide with cellulose-triacetate carbons.

Experimental activation energies E, and pre-exponential factors, A, were obtained by application of the Arrhenius equation, $k = A \exp(-E/RT)$; Fig. 3 shows the Arrhenius plots and Table I includes the parameters obtained from these plots. The Arrhenius plots for both carbons are very similar, having two linear sections with a change in slope at about 1208 K, equivalent to a change in activation energy from 68 to 44 kcal/mole for the 1229 K carbon, and from 69 to 48 kcal/mole for the 1473 K carbon, as the temperature is raised. These results suggest that for both carbons in the temperature range 1160-1208 K, i.e., within Zone 1 of the classification of Hedden and Wicke[13] for reaction of gases with porous carbons, the reaction is chemically controlled. At temperatures above 1208 K there is a transition toward Zone II conditions in which the reaction is controlled by diffusion of reactant and product gases through the pores of the material; values of the pre-exponential factor support this conclusion. Values of experimental activation energy and pre-exponential factor may be compared with those obtained by Rowan[2] for cellulose-triacetate carbon activated by carbon dioxide at atmospheric pressure (Table I); the lower values in the latter case can be attributed to the increased effect of mass transport of gas on the reaction rate at the higher pressure. The values for activation energy found in the present work are comparable with those reported by Bregazzi et al.[14] for reaction of carbon dioxide with poorly outgassed samples of polyvinyl-chloride and polyvinylidene-chloride carbons (70 kcal/mole). In the present case, however, the carbons are thoroughly outgassed before measurements begin.

Adsorption Measurements

Isotherms obtained for the adsorption of carbon dioxide at 195 K on four carbons activated at 1223 K are shown in Fig. 4. The main extent of adsorption occurs at relative pressures below 0.05 for all the carbons, indicating that the major part of the surface of the carbon is contained in very fine pores. The 1229 K unactivated carbon has a greater adsorptive capacity than the 1473 K unactivated carbon, and the carbons activated to 30% burnoff reflect this trend.

Table I. Arrhenius Parameters for Reaction of Cellulose-Triacetate Carbon with Carbon Dioxide

CO_2, torr	Temperature range, K	Experimental activation energy E, kcal/mole	Pre-exponential factor A, $(g \cdot sec^{-1} \cdot g^{-1}) \times 10^5$
		1229 K carbon	
76	1163–1203	68 (±4)	2.3×10^8
76	1203–1233	44 (±3)	4.2×10^3
		1473 K carbon	
76	1163–1213	69 (±4)	8.1×10^7
76	1213–1233	48 (±3)	2.4×10^4
		1233 K carbon*	
760	1163–1203	45 (±4)	5×10^3
760	1203–1233	28 (±3)	11

*Results from S. M. Rowan.[2]

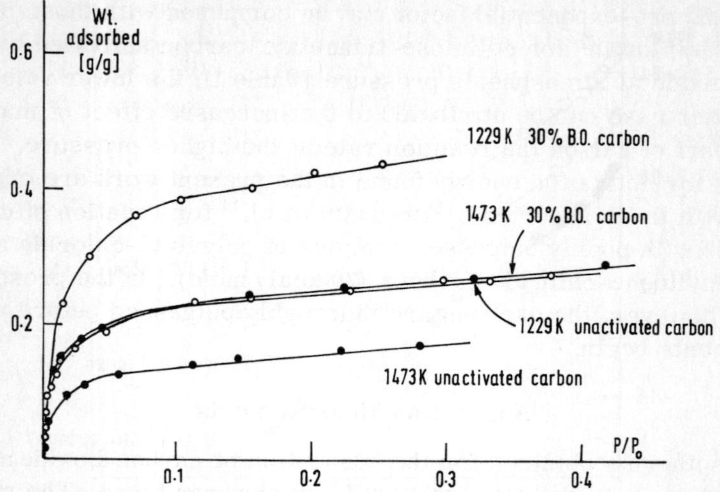

Fig. 4. Adsorption isotherms for carbon dioxide on cellulose-triacetate carbons at 195 K.

The isotherm for the 1473 K carbon activated to 30% burnoff resembles closely that of the 1229 K unactivated carbon, indicating that their microporosity and micropore-size distributions are very similar.

The isotherms have been analyzed by means of the Dubinin equation[15].

$$\log V = \log V_0 - D \log^2 (p_0/p)$$

where V is the volume of adsorbate sorbed at relative pressure p/p_0, V_0 is the micropore volume of the adsorbent, and D is a constant related to the micropore-size distribution. Carbon dioxide is believed to be adsorbed as a supercooled liquid at 195 K,[16] and values of the density (1.23 g · cm^{-3}) and saturated vapor pressure (1413.6 torr) are obtained by extrapolation of data for liquid carbon dioxide to 195 K.[17]

The isotherms plotted according to the Dubinin equation are shown in Fig. 5, and the values of micropore volume and constant D obtained from these are collected in Table II. The micropore volumes for the 1229 K carbons are similar to those obtained for cellu-

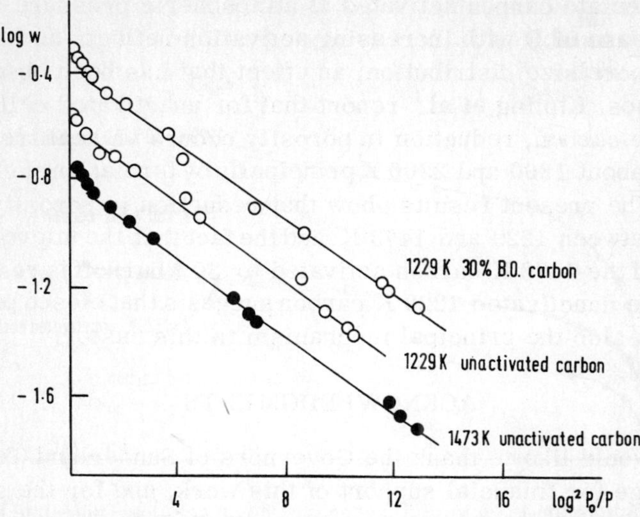

Fig. 5. Dubinin plots of adsorption isotherms for carbon dioxide on cellulose-triacetate carbons. (Note: The plot for the 1473 K, 30% burnoff carbon, which is almost identical to that for the 1229 K unactivated carbon, is omitted.)

Table II. Micropore Volumes for Cellulose-Triacetate Carbons from Application of the Dubinin Equation to Adsorption of Carbon Dioxide at 195 K

Deg. of activation, % burnoff	Temp. of activation, K	Micropore vol., $cm^3 \cdot g^{-1}$	$D \times 10^2$
	1229 K carbon		
0	–	0.21	7.5
30	1223	0.37	8.3
	1473 K carbon		
0	–	0.13	7.3
30	1223	0.22	7.7
	1223 K carbon*		
0	–	0.21	6.4
30	1233	0.38	8.3

* Results from S. M. Rowan.[2]

lose-triacetate carbon activated at atmospheric pressure (Table II).[2] The increase of D with increasing activation reflects an increase in micropore size distribution, an effect that has been observed many times. Kipling et al.[3] report that for unactivated cellulose-triacetate carbon, reduction in porosity occurs on heat treatment between about 1800 and 2300 K principally by formation of closed pores. The present results show that reduction in porosity also occurs between 1229 and 1473 K, and the fact that the micropore volume of the 1473 K carbon activated to 30% burnoff is restored to that of the unactivated 1229 K carbon suggests that closed pore formation is also the principal mechanism in this case.

ACKNOWLEDGMENTS

We would like to thank the Governors of Sunderland Technical College for financial support of this work, and for the grant of a Research Studentship to one of us (N. G. D.).

REFERENCES

1. P. L. Walker, Jr., T. G. Lamond, and J. E. Metcalfe III, Proceedings of the 2nd Conference on Industrial Carbon and Graphite, Society of Chemical Industry, London (1966), p. 7.
2. S. M. Rowan, M. Sc. Thesis, Durham University, England (1966).
3. J. J. Kipling, J. N. Sherwood, P. V. Shooter, and N. R. Thompson, Carbon, 1, 321 (1964).
4. J. J. Kipling and B. McEnaney, Fuel, 43, 367 (1964).
5. J. J. Kipling and P. V. Shooter, Proceedings of the 2nd Conference on Industrial Carbon and Graphite, Society of Chemical Industry, London (1966), p. 15.
6. J. D. Brooks and G. H. Taylor, Carbon, 3, 185 (1965).
7. A. Bailey and D. H. Everett, Nature, 211, 1082 (1966).
8. N. G. Dovaston, B. McEnaney, and S. M. Rowan, Proceedings of 3rd Conference on Industrial Carbon and Graphite, Society of Chemical Industry, London (1970), in press.
9. C. A. Redfarn and J. Bedford, Experimental Plastics, Interscience, London (1960), p. 15.
10. B. McEnaney and S. M. Rowan, Chem. & Ind., 1965, 2032.

11. J. M. Thomas and J. A. Poulis, Vacuum Microbalance Techniques, Vol. 3 (K. M. Behrndt, ed.), Plenum Press, New York (1963), p. 15.
12. M. J. Bennet and F. C. Tompkins, Trans. Faraday Soc., 53, 185 (1957).
13. K. Hedden and E. Wicke, Proceedings of the 3rd Conference on Carbon, Pergamon Press, New York (1959), p. 249.
14. E. Bregazzi, E. Greenhalgh, J. W. Sutherland, and D. J. Tucker, Carbon, 3, 73 (1965).
15. M. M. Dubinin, Chemistry and Physics of Carbon, Vol. 2 (P. L. Walker, Jr., ed.), E. Arnold, London (1966), p. 51.
16. M. M. Dubinin, B. P. Bering, V. V. Serpinski, and B. N. Vasil'ev, Surface Phenomena in Chemistry and Biology (J. F. Danielli, ed.), Pergamon Press, London (1958), p. 172.
17. O. C. Bridgemann, J. Am. Chem. Soc., 49, 1174 (1927).

Microweighing in Vacuo with the Aid of Vibrations of a Thin Band

Th. Gast

Technische Universität
Berlin, W. Germany

If a very thin band of elastic material, e.g., polycarbonate plastic is clamped at both ends, and stretched, the natural frequency of transversal vibration depends on both tension and mass distribution. Let us assume that the band is loaded by a thin layer of dust, formed by spraying or electrostatic precipitation. In this case, the natural frequency will be diminished. This is shown in Fig. 1, where the natural frequency of a plastic band with a thickness of about 3 μ, a width of 20 mm, and a length of some centimeters is shown. This frequency is measured by means of a resonance method for several amounts of precipitated dust. The experiment was carried out in vacuo. Excitation was induced by electrostatic forces and the amplitude of the vibrations was measured by an arrangement similar to a capacitance microphone. By varying the frequency of excitation, the resonance curve could be traced. Without load the peak of the curve was found to be at 1100 Hz, whereas, with a load of 0.2 mg, the peak shifted to 1000 Hz.

This corresponds to a decrease of 10% in frequency. The method, therefore, seems to be practicable and very sensitive. It should be possible to measure variations in mass down to 1 μg. The relation between increase in mass and decrease in frequency is nonlinear. In order to attain a linear relationship between signal and the variations in load, another experiment has been carried out using feedback control to keep the frequency constant. The variations in tensions necessary are indicated (Fig. 2).

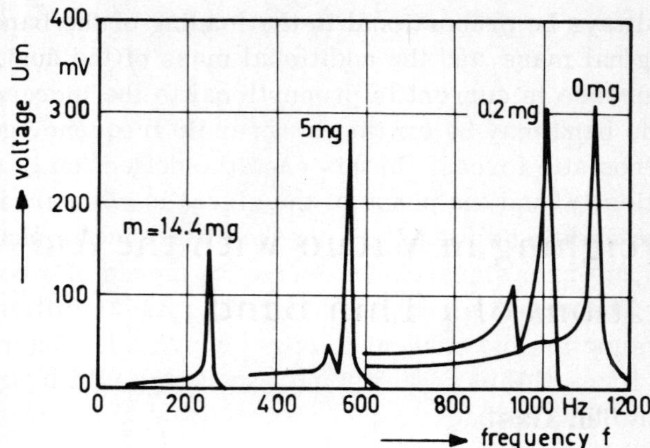

Fig. 1. Resonance curves at different dust masses m and at a pressure of 5×10^{-3} torr.

The band was clamped between a solid block and the moving coil of a dynamic loudspeaker system. When a current was applied to the moving coil, a tension resulted in the band, which was proportional to the current. If, now, the current is controlled in such a manner that the natural frequency of the band remains constant, the tension of the band and, therefore, the current in the moving

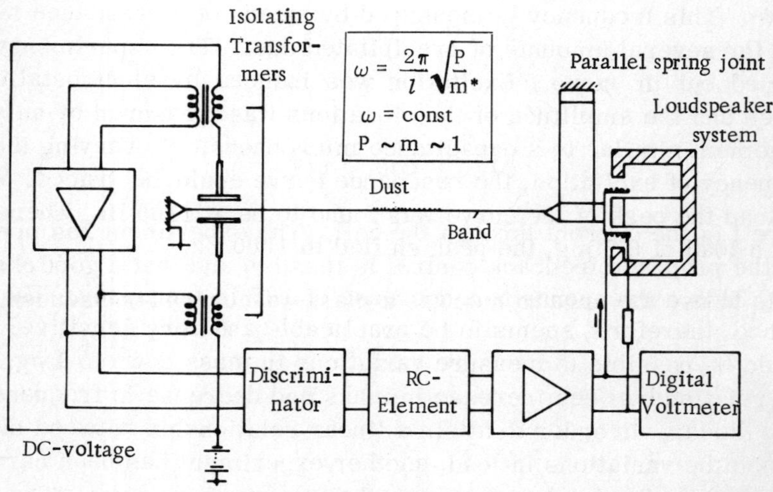

Fig. 2. The frequency controlled mass — current transducer.

coil, will always be proportional to the loading of the band, including the original mass and the additional mass of the dust. Therefore the increase in current is proportional to the increase in weight. The band may be excited by an audio frequency generator, using electrostatic forces. In this case the deflection is measured electrostatically, and the phase of the signal is compared with the phase of the exciting voltage. From the difference in phase of the two signals, an error signal can be derived by means of a controlled rectifier. The error signal is then fed to an amplifier, which in its turn supplies the current to the moving coil. Figure 2 shows a modified form of this arrangement, where the vibrating band is enclosed into the feedback loop of an amplifier and thereby excited to vibrate at its natural frequency.

The audio frequency signal is fed to a frequency discriminator, which derives the necessary error signal.

By means of this closed loop, the shift in frequency will always be minimized. According to the expression

$$\omega = (2\pi/l)\sqrt{P/m^*}$$

where ω is the frequency, P is the tension, m^* is the load per unit length, and l is the length of the band, the ratio P/m^* is kept constant, which means that

$$\Delta P/P = \Delta m^*/m^*$$

or

$$\Delta I/I = \Delta m^*/m^*$$

where I is the current through the coil. The experiment has proved that the proposed feedback control is feasible and that a good chance exists to use the arrangement as a mass-to-current transducer.

Some Uses of Wire-Suspended Microbalances

S. J. Gregg

Brunel University
Acton, London, W. 3, England

ABSTRACT

Processes in which a solid takes up or gives off a gas may be followed by measuring either the change in mass of the solid (gravimetric method) or the change in pressure or volume of the gas (volumetric method). The former method offers a number of advantages, particularly the directness of the measurement, over the latter. These are illustrated in the course of the present paper, which is concerned with the application of a microbalance of the Gulbransen type to the reactions of metals with oxygen and other gases, and of carbon with oxygen, and also with the use of the Cahn recording microbalance for the determination of isotherms of physical and chemical adsorption.

REACTION OF METALS WITH GASES

A monolayer of chemisorbed oxygen weighs $\sim 0.03\,\mu g \cdot cm^{-2}$, so that with the specimens of sheet metal 10 cm^2 in area used in the present experiments, 1 μg corresponds to ~3 monolayers, i.e., to a film of oxide of thickness ~10 Å. The several balances of the Gulbransen type which were used had a sensitivity of ~1 μg, so that the range covered was the "thin-film" region of Mott and Cabrera,[1] say 50 Å $< \tau <$ 500 Å. Pressures between 10^{-4} and 10 torr were avoided because of disturbances resulting from thermomolecular flow, and for the most part pressures of gas around 100 torr were used.

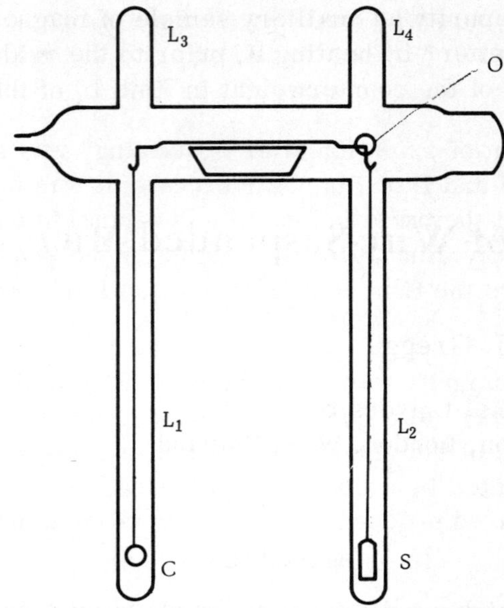

Fig. 1. Microbalance of the Gulbransen type without detachable joints. C) Counterweight; O) optical flat; S) sample.

Reactions between metals and gases can be very sensitive to impurities in the gas, even in trace amounts. The balances were accordingly constructed without demountable joints so as to avoid contamination from grease and through leakage. Loading was accomplished through limbs L_3 and L_4 (Fig. 1) which were sealed off before the run commenced.

The systems which have been studied with balances of this design include the following: magnesium with oxygen[2] or nitrogen[3]; aluminum with oxygen or carbon dioxide[8]; chromium with carbon monoxide or carbon dioxide[9]; zirconium with carbon dioxide[10]; beryllium with oxygen, water, carbon dioxide or carbon monoxide[11]; carbon with oxygen.[4] Some of these are described below by way of illustrating the versatility of the balance.

Oxidation of Magnesium[2]

Various considerations led one to suspect that minute traces of impurity present even in spectroscopically pure oxygen might influence the oxidation of magnesium in oxygen. In order to remove the suspected impurity an auxiliary sample of magnesium was made to serve as a "getter" by heating it, prior to the oxidation run proper, in place of the counterweight in limb L_1 of the balance case.

The amount of oxidation after "gettering" was much less than before (Curves I and II of Fig. 2); moreover it was now "protective" in nature, in that the oxidation came to a virtual halt while the film was still relatively thin (40 Å) in contrast to the behavior in ordinary oxygen where the film continues into the thick-film region ($\gg$500 Å).

The trace impurity was demonstrated to be a hydrocarbon by the following facts: (a) a curve almost identical with curve I resulted if, instead of the magnesium getter, a catalyst for hydrocarbon oxidation heated in a side tube, was used; (b) curve III was obtained if 500 ppm of n-decane was introduced by a break-seal device.

The mechanism of the process cannot be deduced from the gravimetric data alone. Tracer experiments demonstrated that carbon (probably as carbide) is deposited in the film from the hydrocarbon in the gas phase; this film brings about the cracking of the oxide layer so that its protective action is diminished.

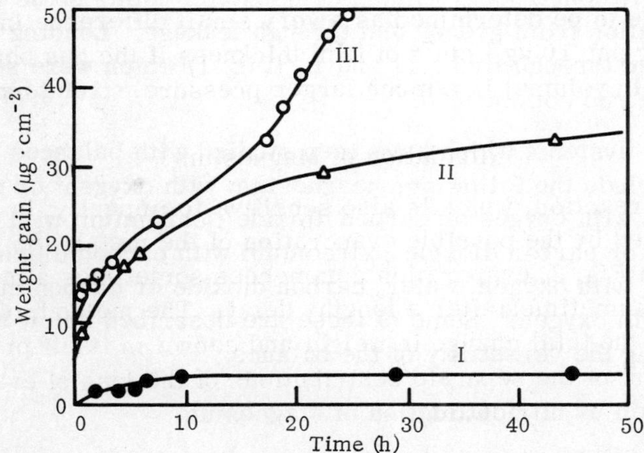

Fig. 2. Oxidation of electropolished magnesium at 525 C in oxygen at 100 torr pressure. I) "Gettered" oxygen; II) "ungettered" oxygen; III) "ungettered" oxygen + 500 ppm n-decane.

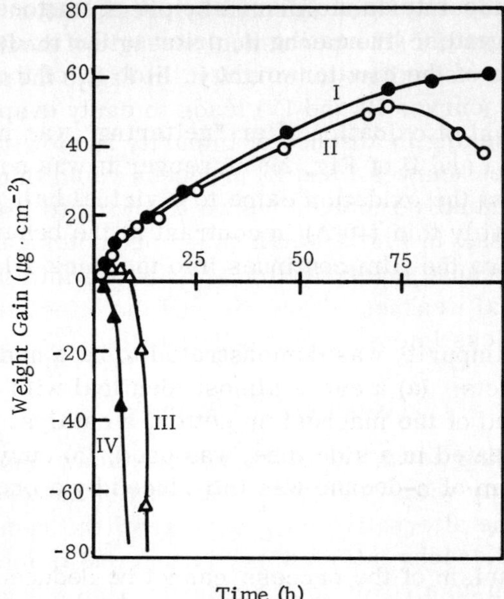

Fig. 3. Nitridation of electropolished magnesium at 500 C. I, II) Commercially pure nitrogen; III, IV) "gettered" nitrogen.

The alternative, volumetric, technique would have offered a much less elegant means of investigation: the uptake of oxygen would have to be determined as a very small difference in pressure (~0.7 torr per 10 $\mu g \cdot cm^{-2}$ of film thickness if the gas phase is 100 cm^{-3} in volume) in a much larger pressure, ~100 torr.

Nitridation of Magnesium[3]

This reaction, which is also sensitive to impurity, is further complicated by the possible evaporation of the metal. As can be seen from Fig. 3, evaporation commences sometimes almost at once and sometimes after a lengthy delay. The method, of course, registers the total change in weight and cannot in itself provide an assessment of the separate contributions of the loss of evaporation and the gain by nitridation.

The actual outcome in a given set of conditions is determined by the interplay of the following factors: (a) the rate of thickening of the nitride film; (b) the rate of growth in size of the cavities at the metal–nitride interface (formed by precipitation of vacancies from the metal); and, (c) the mechanical strength of the film. Factors (a) and (c) are very sensitive to impurity. In Fig. 3 the use of "gettered" nitrogen (curves III and IV) leads to early evaporation because film thickening in absence of impurity is slow (as with MgO); ordinary nitrogen (curves I and II) produces rapid thickening of the film, which eventually ruptures (curve II) owing to its impurity-induced weakness. In Fig. 4 water vapor promotes early evaporation probably because incorporation of hydroxyl into the film produces mechanical weakness (curve I), and the effect of decane is similar though less intense (curves II and III).

Oxidation of Carbon[4]

An outstanding advantage of the microbalance technique is the ease and certainty with which the outgassing of an adsorbent can be monitored; the alternative method, in which the reduction in pressure is followed, is less satisfactory, in that one is looking for very small changes in an already low pressure. This advantage is particularly relevant to studies of the adsorption of water vapor on carbon, since this is very sensitive to the presence of chemisorbed oxygen on the carbon surface.

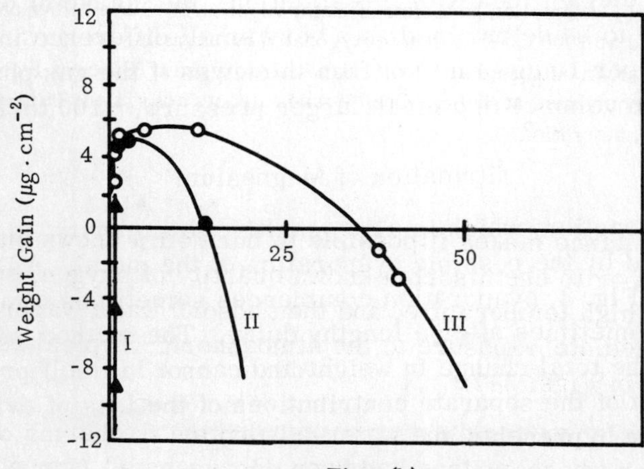

Fig. 4. Nitridation of electropolished magnesium in "gettered" nitrogen at 500 C. I) 500 ppm water vapor added; II, III) 500 ppm n-decane vapor added.

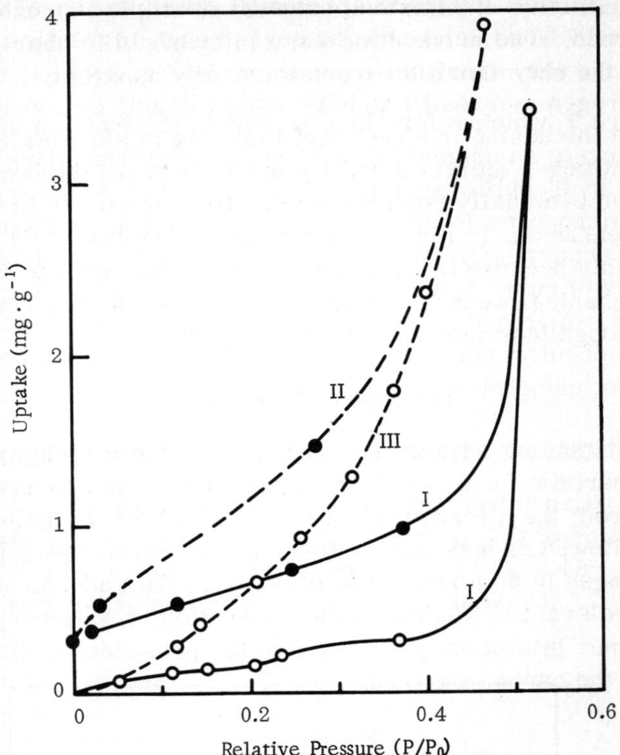

Fig. 5. Adsorption isotherms of water vapor at 25 C on a carbon prepared at 1000 C. I) Carbon deoxygenated by outgassing at 1000 C; II) carbon outgassed at 1000 C, exposed to oxygen at 380 C (cf. Fig. 6), then outgassed at 380 C. Open symbols denote adsorption; solid symbols, desorption.

The balance makes it possible to burn off a known quantity of the carbon or to chemisorb a known quantity of oxygen, outgas the sample at high temperature, and then adsorb water vapor, all without intermediate exposure to the atmosphere. Typical results are presented in Figs. 5–7.

Figure 5 presents the water-adsorption isotherms obtained on a carbon whose surface had been deoxygenated (curve I) by outgassing at 1000 C, and then oxygenated (curve II). Oxygenation was achieved by first outgassing at 1000 C and then exposing to oxygen at 380 C, when 270 μg of oxygen was gradually taken up (Fig. 6) in a form which, since it withstood outgassing at 380 C, must have been chemisorbed. The marked increase in hydrophilic character induced by the chemisorbed oxygen is readily apparent.

Figure 7, which refers to a graphite, illustrates a different point. Despite an outgassing at 1000 C, curves I and II both indicate distinctly hydrophilic behavior. This is probably due to the presence of micropores (width <20 Å) which enhance the adsorption over that occurring in transitional pores (width 20 to 500 Å); the hysteresis of curve I indicates that the pores contain very fine constrictions, which are removed during burnoff prior to curve II. The existence of micropores was confirmed by measurements of butane adsorption, using the same balance.

THE USE OF THE CAHN AUTORECORDING BALANCE

This balance has been used for the determination of isotherms of adsorption of CO_2 on active alumina and active magnesia over the range −84 to 250 C, and also of CO on alumina over the range −196 to 0 C.[5] The ease of monitoring of the outgassing was again

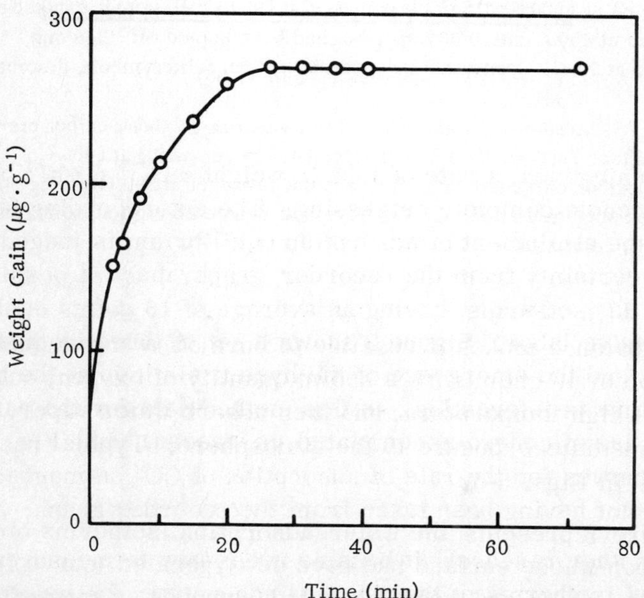

Fig. 6. Rate of chemisorption of oxygen at 380 C by a carbon prepared at 1000 C; oxygen pressure, 76 torr.

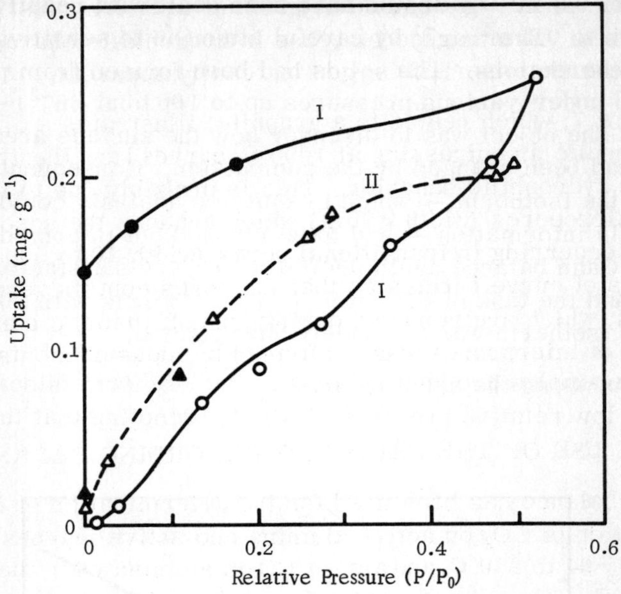

Fig. 7. Adsorption isotherms of water vapor at 25 C on a graphitic carbon prepared at 3000 C. I) Sample outgassed at 1000 C; II) sample exposed to oxygen at 450 C until 0.06% of carbon had been burned off, then outgassed at 25 C. Open symbols denote adsorption; solid symbols, desorption.

very advantageous, a rate of loss in weight $<10^{-5}$ g · h^{-1} being taken to denote complete outgassing. The autorecording feature, because the attainment of adsorption equilibrium is judged rapidly and with certainty from the recorder graph, made it possible to measure 13 isotherms, having an average of 15 points each, without excessive labor. Figure 8 shows a set of three isotherms thus obtained, and the emergence of the hysteresis loop at the highest temperature is interesting. In this method data for the rate of adsorption are of course accumulated en passant, and Fig. 9 gives a set of curves for the rate of adsorption of CO_2 on magnesia, each plotted point having been taken from the recorder graph.

In another research[6] it became necessary to measure a large number of isotherms of the physical adsorption of nitrogen at −195 C on various solids having specific surface areas in the range 3 to 250 $m^2 \cdot g^{-1}$; solids having surface areas up to 400 $m^2 \cdot g^{-1}$, and down to 0.5 $m^2 \cdot g^{-1}$ could have been dealt with readily, and areas down to 0.1 $m^2 \cdot g^{-1}$ by careful attention to sensitivity and to buoyancy corrections. The solids had been formed from powders compacted under various pressures up to 100 tons $\cdot$ in^{-2} ($\sim 1.5 \times 10^9$ $N \cdot m^{-2}$). The object was to discover how the surface area and the porosity had been changed by the compaction. Consequently, the course of the isotherm — which by suitable analysis could be made to yield this information — had to be mapped out in considerable detail. The Cahn balance again proved to be very satisfactory for the purpose, and the task of determining 44 isotherms with 30 to 40 points per isotherm was not unduly burdensome.

Two examples are shown. In one, the isotherm is scarcely altered at low relative pressure (Fig. 10), denoting that the specif-

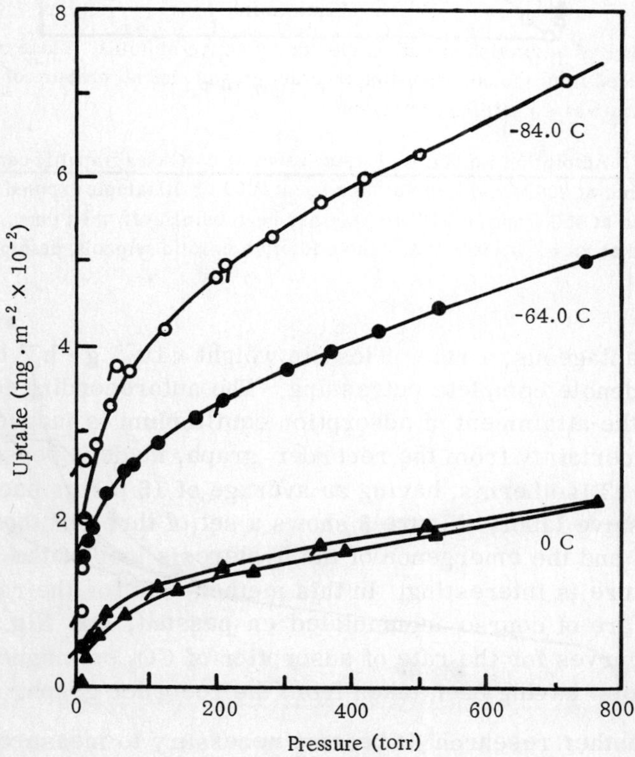

Fig. 8. Adsorption isotherms of carbon monoxide on active alumina, determined by the recording balance. Tails denote desorption.

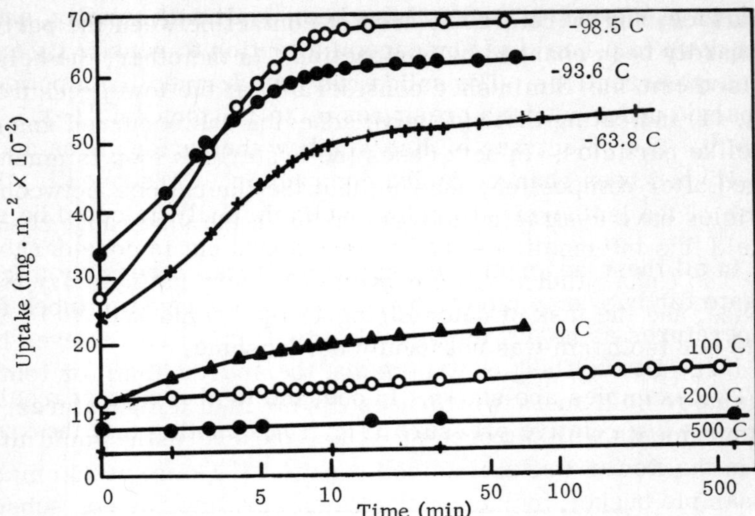

Fig. 9. Rate of adsorption of carbon dioxide by active alumina. These curves were reduced from the corresponding recorder graphs. Initial pressure of carbon dioxide was 15 torr for each graph.

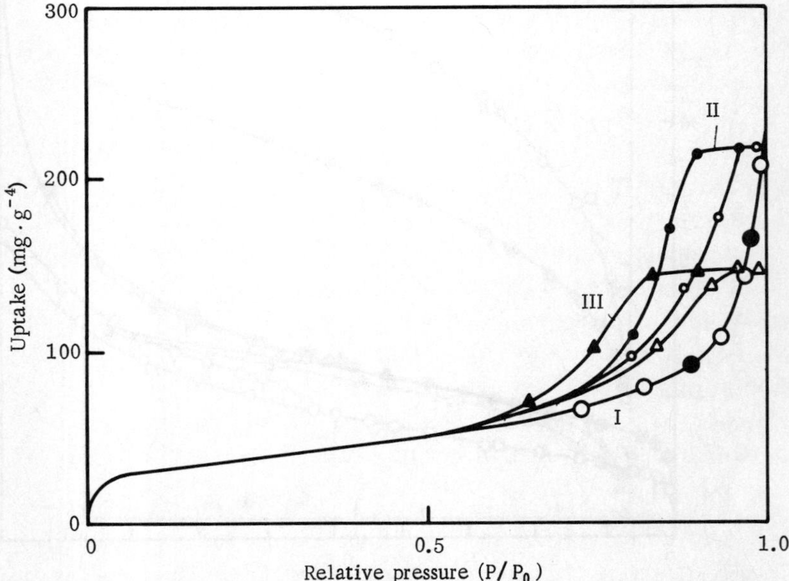

Fig. 10. Adsorption isotherms of nitrogen at −195 C on an alumina composed of spherical particles. I) Loose powder; II) after compaction at 32 tons · in^{-2}; III) after compaction at 96 tons · in^{-2}. Open symbols denote adsorption, solid symbols desorption. Many points are omitted for the sake of clarity.

ic surface, and therefore the area of contact between the particles, had hardly been changed by compaction. In the other, the height of the isotherm has diminished considerably at the low-pressure end (Fig. 11) indicating that extensive adhesion has occurred among the platelike particles. In both cases the hysteresis loop is much enlarged after compaction, showing that the interstices between the particles have contracted to form pores in the size range 20-500 Å.

In all these adsorption experiments it was necessary to compensate for buoyancy effects by keeping both hangdown tubes at temperatures as nearly as possible equal; this was achieved by use of a large Dewar flask containing the thermostat fluid for temperatures up to 60 C, and a twin furnace for the high temperatures. In measurements with nitrogen at −196 C the level of the liquid nitrogen in the Dewar was maintained constant (±1 mm) at 25 cm above the sample bucket, by means of an auto-leveling device; subsequent work by Cutting reported in another paper at this conference

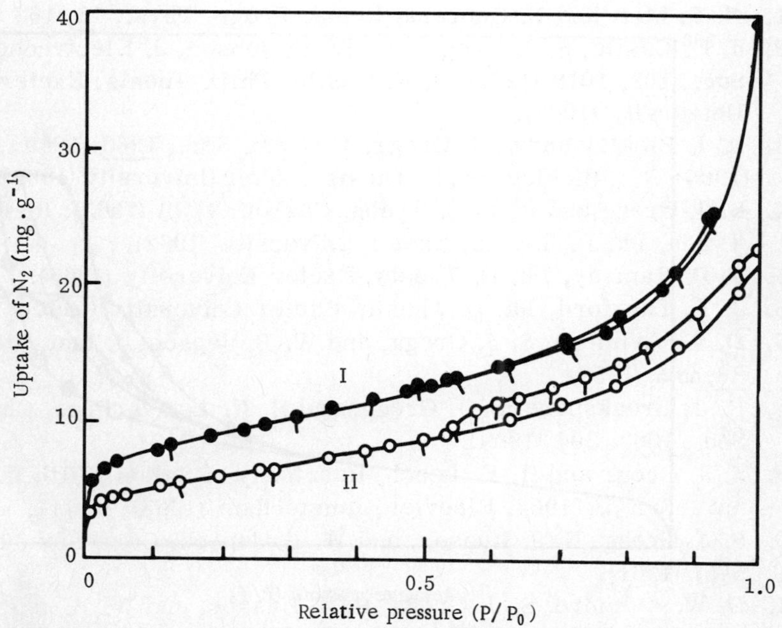

Fig. 11. Adsorption isotherms of nitrogen at −195 C on finely divided mica. I) Loose powder; II) after compaction at 96 tons · in^{-2}. Tails denote desorption.

suggests that the temperature of the sample must nevertheless have been higher, probably by 0.5 deg, than the surrounding bath.

CONCLUSION

Vacuum microbalances of the wire-suspended type provide a suitable means of measuring isotherms of physical and chemical adsorption, and for studying gas–solid reactions, such as the oxidation of metals by oxygen and other gases. Particularly useful features are the speed and directness with which the uptake or loss of gas is measured, the ease and certainty of monitoring of the outgassing, the ready possibility of studying the effect of trace impurities in the gas, and (with a recording balance) the emergence of kinetic data. For elucidation of the mechanism of a gas–solid reaction, however, the balance — since it measures the change in mass, but not in nature, of the solid — may need to be supplemented by other techniques, such as the use of tracers.

REFERENCES

1. N. F. Mott and N. Cabrera, Repts. Progr. Phys., 12, 163 (1948).
2. J. E. Castle, S. J. Gregg, and W. B. Jepson, J. Electrochem. Soc., 109, 1018 (1962); J. E. Castle, Ph.D. Thesis, Exeter University (1961).
3. R. I. Bickley and S. J. Gregg, J. Chem. Soc., 1966, 1849 (1966); R. I. Bickley, Ph. D. Thesis, Exeter University (1962).
4. S. J. Gregg and R. F. S. Tyson, Carbon, 3, 39 (1965); R. F. S. Tyson, Ph. D. Thesis, Exeter University (1962).
5. J. D. Ramsay, Ph. D. Thesis, Exeter University (1965).
6. J. F. Langford, Ph. D. Thesis, Exeter University (1967).
7. D. W. Aylmore, S. J. Gregg, and W. B. Jepson, J. Inst. Metals, 88, 205 (1959).
8. R. J. Breakspere, S. J. Gregg, and H. F. Leach, Proc. Chem. Soc., 1963, 304 (1963).
9. S. J. Gregg and H. F. Leach, Reactivity of Solids, Fifth Internatl. Symp., 1964, Elsevier, Amsterdam (1965), p. 311.
10. S. J. Gregg, R. J. Hussey, and W. B. Jepson, Corrosion, 17, 575t (1961).
11. D. W. Aylmore, S. J. Gregg, R. J. Hussey, and W. B. Jepson, J. Nuclear Materials, 2, 1969 (1960); 3, 175, 190 (1961); 4, 46 (1961).

Oxidation in Flow-Reaction Systems*

Earl A. Gulbransen, Fred A. Brassart,

and Kenneth F. Andrew

Westinghouse Research Laboratories
Pittsburgh, Pennsylvania 15235, U. S. A.

ABSTRACT

Many high-temperature metals form volatile oxides when volatilized above 1200 C. When that occurs, rapid oxidation is observed. Special dynamic vacuum-microbalance reaction systems are used to study the several chemical and physical mechanisms involved. To illustrate the method, the oxidation of rhenium and molybdenum is presented. Rates of oxidation of 10^{18} atoms of Re or Mo were observed at 1100 and 1400 C, respectively. Comparison was made to predictions from the activated-state theory of surface reactions.

INTRODUCTION

At the Pittsburgh Vacuum Microbalance Conference in 1964 a vacuum microbalance and flow-type reaction system for studying rapid oxidation reactions was described as applied to a study of the combustion of graphite.[1] Earlier studies on the combusion of graphite[2] in which a static reaction system was used showed that the reaction system itself limited the magnitude of the rates of reaction which could be measured. By imposing gas flow over the graphite sample and by decreasing the geometrical surface area of the sample, rate measurements for the reaction at 9 torr and 1322 C could be extended to 10^{20} atoms of C reaction $\cdot$ cm^{-2} $\cdot$ sec^{-1}. This

* The work described here was supported in part by the U. S. Army Research Office, Durham, North Carolina, under Contract No. DA-31-124-ARO-D-196.

rate is equivalent to about 10^5 Å of C reacting per second when calculated on the basis of geometrical area. Sample porosity and the possibility of spalling with graphite samples make a complete theoretical interpretation of these fast reactions difficult.

This paper presents a study on the fast oxidation reactions of rhenium and molybdenum under conditions where volatile oxides form. Rhenium and molybdenum are dense, nonporous metals. Reaction rates for these metals can be compared directly to theory.

THEORY

Four types of oxidation processes occur when solids that form volatile oxides are oxidized. These are:

(1) at low temperatures, transport of metal and oxygen atoms or ions through the solid or liquid oxide film or scale on the metal;

(2) at higher temperatures, surface chemical reactions of oxygen molecules or atoms at the nearly oxide-free metal—gas interface;

(3) at still higher temperatures, diffusion of oxygen gas through a barrier layer of volatilized oxide to the solid—oxygen reaction interface;

(4) finally, at high rates of reaction, spalling of small crystallites from the solid.

Process (1) can be distinguished from processes (2), (3), and (4) by the sign of the weight change and an analysis of the weight change and oxygen consumed during oxidation. In the absence of spalling, processes (2) and (3) can be distinguished from each other by the nature of the temperature and pressure dependence of the rate of oxidation. Processes (2) and (3) can be distinguished from process (4) by visual observations on the specimen.

Surface chemical reactions include the several types of adsorption and desorption processes. They can be distinguished by the magnitude of the enthalpy of activation and the effect of pressure on the kinetics of reaction.

Eyring and coworkers[3] developed rate expressions for several types of surface reactions, including those where the rate-controlling step is mobile adsorption, chemical reaction, or desorption.

Their expression for mobile adsorption is given by

$$\frac{dn'}{dt} = \frac{S}{l} C_g kT (2\pi mkT)^{-\frac{1}{2}} \exp\left(\frac{-\Delta H_1}{RT}\right) \qquad (1)$$

where dn'/dt is the rate of reaction in atoms of Re $cm^{-2} \cdot sec^{-1}$; l is the stoichiometric factor; S is the surface area (cm^2); C_g is the concentration of O_2 molecules (cm^{-3}) in the gas phase; k is Boltzmann's constant; T is the absolute temperature; m is the mass of the O_2 molecules; R is the gas constant; and ΔH_1 is the enthalpy of activation of adsorption. Earlier studies on the oxidation of carbon[2] and of molybdenum[4] have shown that mobile adsorption is probably the rate-controlling step.

To calculate the true rate of reaction for porous materials or rough surfaces, S must be determined by gas-adsorption methods. For graphite, S can have a value of 1000 cm^2 or higher.

Another likelihood is that the rate-controlling step is the desorption of the oxide. The activated-state-theory expression for desorption[3] is

$$\frac{dn'}{dt} = \frac{S}{l} C_a \frac{h}{kT} \exp\left(\frac{-\Delta H_2}{RT}\right) \qquad (2)$$

where C_a is the number of reaction sites per cm^2; ΔH_2 is the enthalpy of activation of desorption; and h is Planck's constant.

According to the activated-state theory of surface reactions, a zero-order chemical surface reaction would give the same equation as that for desorption. To apply these equations it is necessary to determine dn'/dt, ΔH, and the pressure dependence of dn'/dt.

The purpose of comparing theoretical predictions to experimental data is to determine the factors that limit the use of metals in high-temperature oxidizing environments. From the experimental side, careful consideration must be given to the effects of possible side reactions of the specimen support, furnace tube, and thermocouple assembly with the specimen and the oxidizing atmosphere.

Table I summarizes some of the factors which limit the usefulness of a metal or solid in an oxidizing environment at high temperature. Few of the elements have melting points greater than 2000 C, and these form volatile oxides. When volatile oxides form,

Table I. Limiting Factors for Oxidizing High-Temperature Materials

1. Melting point of material
2. Melting point of oxide
3. Vapor pressure of material, its oxide, or both
4. Solubility of oxide in metal
5. Transport of oxygen through oxide layer
6. Spalling of oxide
7. Chemical reaction processes at oxide free surface
8. Gas phase diffusion of oxygen to reaction zone

the protective effect of the oxide film is lowered and rapid oxidation occurs.

Table II shows a compilation of the melting and boiling points of six elements, each having a melting point above 2000 C. The temperatures at which volatile oxides are observed experimentally and the oxidation equations are given. The data show volatile oxides forming at relatively low temperatures for C, Re, and Mo.

LITERATURE

Lavrenko[5] studied the oxidation of 99.94% Re over the temperature range of 350-750 C. Linear oxidation rates were observed. In a recent paper, Phillips[6] has reviewed some of the recent work and has extended the oxidation studies to 1300 C. Linear oxidation rates were observed for air oxidation. The oxide formed was Re_2O_7. Gulbransen and Brassart[7] studied the oxidation of rhenium and a rhenium — 8%-titanium alloy in flow environments. That work will be discussed further in this paper. A literature survey on the oxidation of molybdenum was given in an earlier paper.[4]

EXPERIMENTAL

Rhenium and molybdenum are the best metals for studying high-temperature oxidation processes when volatile oxides form, since these oxides show appreciable volatility below 1000 C. The oxide Re_2O_7 melts at 296 C and boils at 362 C. MoO_3 melts at 795 C and boils at 1155 C.

A flow-reaction system[1] was used for all kinetic measurements. The specimens were right cylinders having surface areas of about 0.75 cm^2, and were suspended from one end of a sensitive recording electrobalance. The balance had a sensitivity of 10^{-7} g · mm^{-1} deflection on the recorder chart, a period of less than 0.1 sec, a load capacity of 1 g, and a range of 1 g. Reliable weight-change readings were obtained from the recorder chart after 10-30 seconds of elapsed time. The oxygen gas was preheated, led to the bottom of the furnace tube, and exhausted at an outlet at the top of the furnace tube assembly by means of a vacuum pump. A servo-operated valve was incorporated into the gas handling system to automatically control pressure and gas flow. Pressures of 1 to 10

Table II. Oxidation Reactions, Elements Forming Volatile Oxides

Element	M.p., C	B.p., C	Temp. for observing volatile oxide, C	Reaction equation
C	3652[a]	4200	> 425	$C(s) + \frac{1}{2}O_2(g) \rightleftharpoons CO(g)$
Re	3180 ± 20	5627[b]	> 600	$2Re(s) + \frac{7}{2}O_2(g) \rightleftharpoons Re_2O_7(g)$
Mo	2620	4507	> 800	$Mo(s) + \frac{3}{2}O_2(g) \rightleftharpoons \frac{1}{3}(MoO_3)_3(g)$
W	3410	5900	>1200	$W(s) + \frac{3}{2}O_2(g) \rightleftharpoons \frac{1}{3}(WO_3)_3(g)$
Nb	2415	3300	>1700	$Nb(s) + \frac{1}{2}O_2(g) \rightleftharpoons NbO(g)$
Ta	2996	6000[b]	>2200	$Ta(s) + \frac{1}{2}O_2(g) \rightleftharpoons TaO(g)$

[a]Sublimation.
[b]Estimated.

torr were used. Gas was added manually; and when the desired pressure was achieved, the reaction system was switched to automatic control. The gas flow was calculated from pressure readings on a calibrated gas-storage vessel.

The formation of films of nonvolatile impurities on the surface can be minimized by using high-purity metal and high-purity oxygen. Reagent-grade oxygen having 8.4 ppm N_2, 20 ppm CH_4, and 15.2 ppm Kr was used. The rhenium was obtained from Materials Research Corp. and had a purity of 99.99%. The molybdenum was obtained from Westinghouse Electric Corporation Lamp Division and had a purity of 99.95%. Spectroscopic analysis showed the following impurities in ppm: Cu, 10; Cr, 45; Mn, 5; Al, 60; Fe, 200; Ca, 10; Ni, 70; Sn, 10; Mg, 5; Si, 50; and B, 1. To avoid accumulation of surface oxides, of impurity atoms, and to avoid surface-area changes, the samples were not reacted extensively.

RESULTS AND DISCUSSION

Rhenium

The oxidation of high-purity rhenium was studied at 1, 2, 5, and 10 torr oxygen pressure for the temperature range of 600-1400 C, and for flow rates of 0.42 to 1.92×10^{19} O_2 molecules per sec. Linear weight-loss vs. time plots were obtained at all temperatures. No evidence was found for a weight gain which would suggest the formation of an oxide film. If an initial oxide film was formed, the weight change involved was negligible compared to the weight losses due to the formation of a volatile oxide. Rates of oxidation, dn'/dt were calculated as atoms of Re reacting $\cdot$ cm^{-2} $\cdot$ sec^{-1}.

Figure 1 shows a log dn'/dt vs. 1/T plot for the kinetic data on the oxidation of pure rhenium. Parallel straight lines were obtained for 1, 2, 5, and 10 torr pressures below 1000 or 1100 C. For that region an enthalpy of activation of 16.7 kcal $\cdot$ $mole^{-1}$ was obtained. Above 1100 C, a break occurs in the plot for 2, 5, and 10 torr oxygen pressures. From a comparison with studies on the combustion of carbon,[2] the change in slope of the log dn'/dt vs. 1/T plot can be considered as a transition between a chemical oxidation process and external gas diffusion of oxygen to the reaction interface. For rhenium, this transition occurs at about 1100 C. The

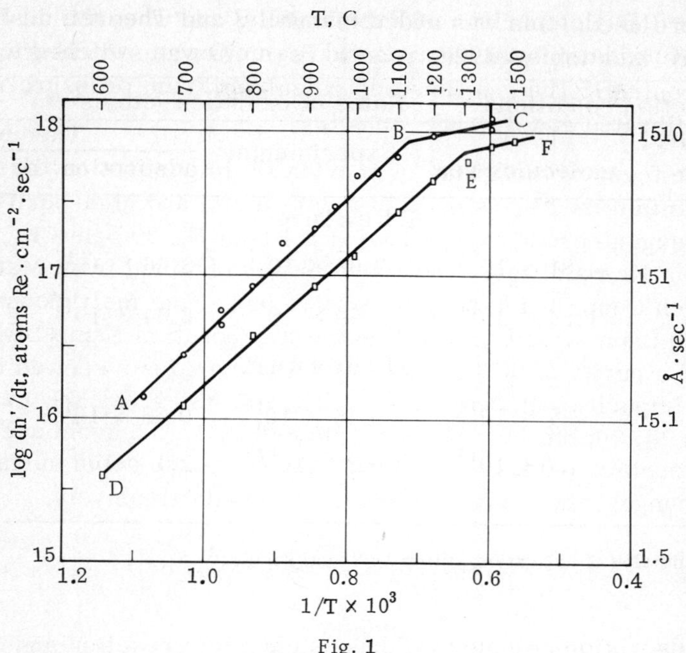

Fig. 1

external gas-diffusion process is probably the diffusion of oxygen molecules through the volatilized oxide to the reaction interface. The temperature coefficient for this region is small.

Equation (1) for mobile adsorption of oxygen molecules can be fitted to the data of Fig. 1 for oxygen pressures of 2, 5, and 10 torr using $\Delta H = 16.7$ kcal/mole. Table III shows a comparison of the experimental rates of oxidation at 1000 C with values calculated from the mobile-adsorption model and the desorption model of chemical processes. Agreement with the mobile adsorption model is good if a surface area of 1 or 2 cm^2 for each cm^2 of geometrical area is used. The pressure dependence at 1000 C is $P^{0.57}$. Although mobile adsorption fits the experimental data, a closer analysis of the pressure dependence indicates additional factors are affecting the rates of oxidation.

In the Table III comparison, it was assumed that all the gas was involved in mobile adsorption, with an activation energy of 16.7 kcal · $mole^{-1}$ for pure rhenium. Actually it is postulated that a monolayer of oxygen is first adsorbed with a lower activation energy and that mobile adsorption of oxygen occurs on that surface.

Table III. Comparison of Experimental and Theoretical Rates of Oxidation at 1000 C, 2 and 5 torr Oxygen Pressure

Reaction conditions		Rates of oxidation (atoms·cm^{-2}·sec^{-1})		
Pressure, torr	Flow, molecules·sec^{-1}	Experimental	Theory adsorption*	Theory desorption*
		Rhenium		
2	4.81×10^{18}	3.08×10^{17}	2.66×10^{17}	2.0×10^{25}
5	1.84×10^{19}	5.53×10^{17}	6.64×10^{17}	2.0×10^{25}
		Molybdenum		
2	2.7×10^{18}	0.46×10^{17}	0.86×10^{17}	2.0×10^{25}
5	2.7×10^{18}	0.8×10^{17}	2.1×10^{17}	2.0×10^{25}

* Basis enthalpy of activation, $\Delta H = 16.7$ kcal · mole^{-1}.

That adsorption complex undergoes chemical reaction and desorption to form Re_2O_7.

It is concluded that the mechanism of oxidation of rhenium for the region AB of Fig. 1 is probably mobile adsorption of oxygen modified by structural changes at the reaction interface. Studies of the reaction interface would be necessary to establish a complete mechanism.

Molybdenum

Nearly linear weight-loss vs. time curves for the oxidation of molybdenum over the temperature range of 800 to 1500 C were obtained. Figure 2 shows a log rate vs. 1/T plot of the rate data, in units of atoms of molybdenum reacted per cm^2 per sec and in Å · sec^{-1}, for reactions at 2 and 5 torr oxygen pressure. Straight lines could be drawn through the experimental data. An enthalpy of activation of 20 kcal · mole^{-1} was calculated. That value is a little higher than the value found for the oxidation of rhenium, but it is in agreement with earlier studies on molybdenum in a static reaction system at higher oxygen pressures.[4]

The data of Fig. 2, and additional data taken at higher oxygen pressures,[4] suggest that a transition is occurring from chemical

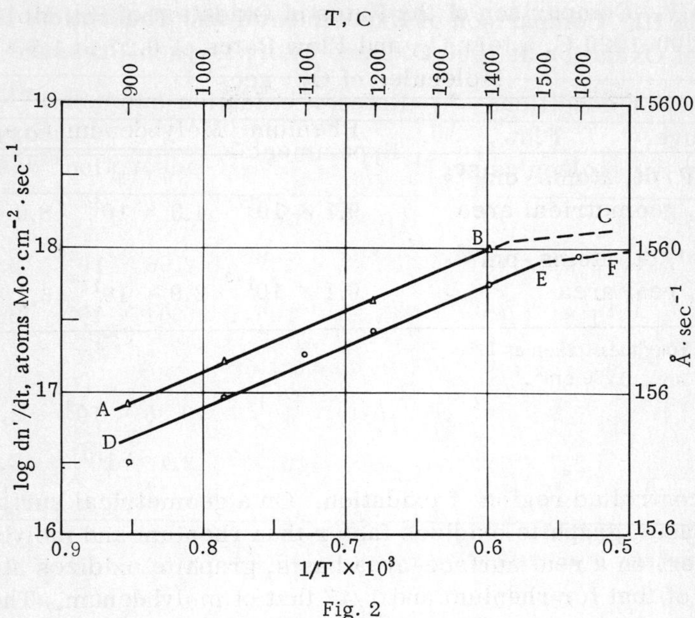

Fig. 2

processes to gas-diffusion-controlled processes for oxidation rates above about 10^{18} atoms of Mo per cm^2 per sec.

Table III shows a comparison of the experimental rates at 1000 C with rates calculated from the mobile-adsorption model and the desorption model of chemical processes by using an enthalpy of activation of 20 kcal · $mole^{-1}$. Agreement with experimental data is good considering the impurity level of the molybdenum and the probable error in evaluating ΔH. Desorption is not a probable rate-limiting step in the mechanism due to the low values found for ΔH and the nearly linear pressure dependence.

The conclusion reached in the kinetic studies on molybdenum[4] was the same for reaction systems of both the flow and static types. The rate-controlling step is mobile adsorption of oxygen molecules on a molybdenum surface covered by a previously adsorbed oxygen monolayer.

Comparison Oxidation of Rhenium, Molybdenum, and Graphite

Table IV shows a comparison of the rates of oxidation of Re, Mo, and C at 1200-1220 C and 5 torr oxygen pressure for the chemi-

Table IV. Comparison of the Rates of Oxidation of Re, Mo, and C at 1200-1220 C, 5 torr O_2, and Flow Rates of 0.27 to 1.9×10^{19} Molecules of $O_2 \cdot sec^{-1}$

	Rhenium	Molybdenum	Graphite
Rate dn'/dt, atoms $\cdot$ $cm^{-2} \cdot sec^{-1}$, geometrical area	9.1×10^{17}	4.3×10^{17}	8.0×10^{18}
Rate dn'/dt, atoms $\cdot$ $cm^{-2} \cdot sec^{-1}$, real area	6.1×10^{17}[a]	2.9×10^{17}[a]	6.2×10^{15}[b]

[a]Surface roughness taken as 1.5.
[b]Surface area, 1290 cm^2.[8]

cally controlled region of oxidation. On a geometrical surface-area basis, graphite oxidizes faster than rhenium and molybdenum.[8] However, on a real surface-area basis, graphite oxidizes at a rate 1/100 of that for rhenium and 1/47 that of molybdenum. The real surface area for graphite was taken from earlier studies.[8] The surface roughness of rhenium and molybdenum was taken as 1.5. The combusion of graphite follows a chemically controlled mechanism at higher total rates of oxidation than do those of rhenium and molybdenum. This is probably due to the smaller molecules associated with the combustion of graphite, since these molecules, CO and CO_2, have less influence on the diffusion of oxygen to the reaction interface.

Comparison of Figs. 1 and 2 for Re and Mo shows the conditions for transition from chemically controlled to gas-diffusion-controlled oxidation. Rhenium and molybdenum show transition at reaction rates of about 1×10^{18} atoms $\cdot$ $cm^{-2} \cdot sec^{-1}$. Graphite[9] shows chemically controlled rate processes to at least 5×10^{19} atoms $\cdot$ $cm^{-2} \cdot sec^{-1}$ for a pressure of 5 torr. Molybdenum showed the lowest transition temperature of about 1100 C, while rhenium showed a transition temperature of about 1400 C.

Efficiency Calculations

Two types of efficiencies are used to compare the experimental oxidation rates with kinetic theory and with the actual flow rate of oxygen molecules and atoms over the sample.

The percentage kinetic theory efficiency ε is given by

$$\varepsilon = (dn'/dt)\, l\, (1/q) \times 100 \qquad (3)$$

Here l, the stoichiometric factor, is $2/3$ for the formation of MoO_3 and 1.75 for the formation of Re_2O_7. The value of q, the number of collisions $\cdot$ $cm^{-2} \cdot sec^{-1}$, is obtained from the equation[10]

$$q = 3.52 \times 10^{22}\, [P/(MT)^{\frac{1}{2}}] \qquad (4)$$

where M is the molecular weight of the gas, T is the absolute temperature, and P is the pressure in torr.

The second efficiency compares the experimental rate of reaction with the rate of gas flow over the sample. Percent flow efficiencies E are calculated from the equation

$$E = l\, (dn'/dt)\, (A/F) \times 100 \qquad (5)$$

where A is the geometric surface area in cm^2, F is the flow rate in molecules of O_2 per sec, dn'/dt is the rate of reaction in terms of atoms of metal $\cdot$ $cm^{-2} \cdot sec^{-1}$, and l is the stoichiometric ratio for the given oxide.

Flow- and kinetic-theory efficiencies in the gas-diffusion region of reaction for the oxidation of rhenium and molybdenum are given in Table V.

Table V. Flow- and Kinetic-Theory Efficiency Calculations

Reaction conditions			Efficiencies	
Temp., C	Pressure, torr	Flow, O_2 molecules$\cdot sec^{-1}$	Flow, E%	Kinetic theory, ε%
		Rhenium		
1400	1	1.39×10^{19}	7.6	0.88
1400	2	4.98×10^{18}	18.6	0.40
1400	5	1.85×10^{19}	6.3	0.24
1400	10	1.91×10^{19}	6.6	0.14
		Molybdenum		
1400	2	2.7×10^{18}	19.1	0.28
1400	5	4.0×10^{18}	21.5	0.19

CONCLUSIONS

Methods using simple laboratory apparatus are available for studying fast oxidation reactions in flow environments under severe oxidizing conditions. Fast oxidation reactions are predicted when thick oxide films are absent due to volatilization. Kinetic studies on rhenium and molybdenum show fast oxidation reactions do occur. Rates of oxidation of 0.19 to 0.40% of those predicted by collision theory are found to occur at temperatures as low as 1400 C for pressures of 1 and 5 torr.

Mobile adsorption of oxygen is suggested as a mechanism of oxidation for rhenium and molybdenum for rates of oxidation below 9×10^{17} atoms of metal reacting $\cdot$ cm^{-2} $\cdot$ sec^{-1} at pressures of 2 to 10 torr. Above that rate of oxidation, reaction is limited by gas diffusion of oxygen through a barrier layer of volatilized reaction products.

For fast flow rates, chemical processes can be extended to higher rates of oxidation and higher temperatures. The chemical rate-controlling enthalpy barriers are not very helpful in preventing fast reactions. For rhenium and molybdenum, those enthalpy barriers are 16.7 and 20 kcal $\cdot$ $mole^{-1}$, respectively. For graphite, an enthalpy barrier of 39 kcal $\cdot$ $mole^{-1}$ was found in an earlier study. Since mobile adsorption of oxygen is probably rate controlling, the magnitude of the enthalpy barrier, $\exp(-\Delta H/RT)$ is the only effective factor separating the observed reaction rate from that predicted from collision theory. Although graphite has the larger enthalpy barrier, the large surface area or porosity of the sample makes the observed reaction rate, on a geometrical-area basis, greater than that for rhenium and molybdenum.

At temperatures above 1400 C, oxygen molecules tend to dissociate to oxygen atoms, and conditions for protection against oxidation become poorer.

To a limited extent, oxide coatings may stabilize oxide films on high-temperature metals. They must be chosen carefully to achieve protection over the severe oxidizing conditions.

It is very difficult to overcome the inherent chemical reactivity of high-temperature materials in oxidizing atmospheres. One area of study which needs to be investigated is the structure of the reaction interface which exists on the metal surface during reaction. High-temperature electron diffraction is one of the promising methods for studying that interface.

REFERENCES

1. E. A. Gulbransen, K. F. Andrew, and F. A. Brassart, Studies of fast reactions in flow environments, in: Vacuum Microbalance Techniques, Vol. 4 (P. M. Waters, ed.), Plenum Press, New York (1965), p. 127.
2. E. A. Gulbransen, K. F. Andrew, and F. A. Brassart, The oxidation of graphite at temperatures of 600° to 1500°C and at pressures of 2 to 76 torr of oxygen, J. Electrochem. Soc., 110, 476 (1963).
3. S. Glasstone, K. J. Laidler, and H. Eyring, Theory of Rate Processes, McGraw-Hill Book Co., New York (1941).
4. E. A. Gulbransen, K. F. Andrew, and F. A. Brassart, Oxidation of molybdenum at 550°-1700°C, J. Electrochem. Soc., 110, 952 (1963).
5. V. O. Lavrenko, Kinetics and mechanism of high-temperature oxidations of rhenium in the recrystallized and cold-worked states, Dopovidi Akad. Nauk RSR, 1216 (1958).
6. W. L. Phillips, Jr., The rate of oxidation of rhenium at elevated temperatures in air, J. Less Common Metals, 5, 97 (1963).
7. E. A. Gulbransen and F. A. Brassart, Oxidation of rhenium and a rhenium — 8%-titanium alloy in flow environments at oxygen pressures of 1-10 torr and at 800-1400°, J. Less Common Metals, 14, 217 (1968).
8. E. A. Gulbransen, K. F. Andrew, and F. A. Brassart, Ablation of graphite in oxygen and air at 1000-1400°C under flow conditions, Carbon, 1, 413 (1964).
9. E. A. Gulbransen and K. F. Andrew, Reactions of artificial graphite, Ind. Eng. Chem., 44, 1034 (1952).
10. J. H. de Boer, The Dynamical Character of Adsorption, Oxford University Press, London (1953).

UHV Microbalance and Quartz Oscillator at Low Temperatures

D. Hillecke* and H. Mayer

Department of Physics
Technische Universität Clausthal
W. Germany

ABSTRACT

The intensities of beams of rubidium atoms were determined simultaneously by two independent methods with a magnetically compensated microbalance and with a resonating quartz oscillator, the latter serving as the pan of the microbalance. Both systems were operating at 180 K inside a cooled glass Dewar. Taking into account the temperature changes of the weighing ensemble during deposition caused mainly by the condensation of Rb atoms and the warming up of the surrounding Dewar, the Rb atom beam intensities, determined by the microbalance and independently by the frequency shift of the resonating quartz, agree to within 2%. In a separate experimental setup the measurements of the Rb atom beam intensities with a calibrated Pt-ionization foil and with the quartz oscillator at 81 K agree also to within 2%.

INTRODUCTION

The first experimental verification of microweighing using the resonance frequency shift of AT-cut quartz crystal oscillators as a result of additional mass loading was given by Sauerbrey[1] and

* Now at Veith-Pirelli A. G., Höchst/O. D. W., W. Germany.

Lostis[2] in 1959. Since that time many different applications of quartz oscillators for microweighing have been demonstrated.[3-13]

THEORY

The theory of the AT-cut quartz oscillator shows that a frequency shift

$$\Delta f = -(f_0/d_0\rho_Q)(\Delta m/F) \tag{1}$$

results, when a mass thickness ($\Delta m/F$) is condensed on the quartz. The theoretical sensitivity

$$C_{theor} = f_0/d_0\rho_Q \tag{2}$$

can be determined very easily by simple measurement of the frequency f_0, the thickness d_0, and the density ρ_Q of the quartz oscillator. Sauerbrey[14] has shown that the area F, on which Δm should be deposited, is determined by the area of the electrodes on the quartz oscillator, as oscillations outside this area decay very rapidly.

EXPERIMENTAL SETUPS AND RESULTS

The determination of rubidium atom beam intensities with an oscillating quartz crystal requires that certain conditions are fulfilled, i.e., the condensation coefficient is unity, and no re-evaporation and no surface diffusion of Rb atoms take place. These can be achieved completely by cooling down the oscillator target.

The ultrahigh-frequency vibration (UHV) experimental setup of the microbalance and the quartz oscillator is shown in Fig. 1. The attachment of the quartz-crystal plate to the microbalance has been achieved by means of the platinum-foil edges of the pan and by gluing with silver cement. The attachment zones are the facettes, so that, according to Sauerbrey,[14] there is no distortion of the quality of the vibrations. Ultraclean rubidium[18] is evaporated from a Rb-atom beam oven onto the cooled pan of the tungsten wire torsion microbalance.* The pan has a hole through which part of the vapor can proceed to deposit on the 5.5-MHz quartz oscillator, serving as part of the pan. As both targets are near together (~4 mm apart) and far away from the Rb oven (~25 cm), both systems measure the same condensed mass thickness

* For details of construction and balance suspension see references 15 and 16.

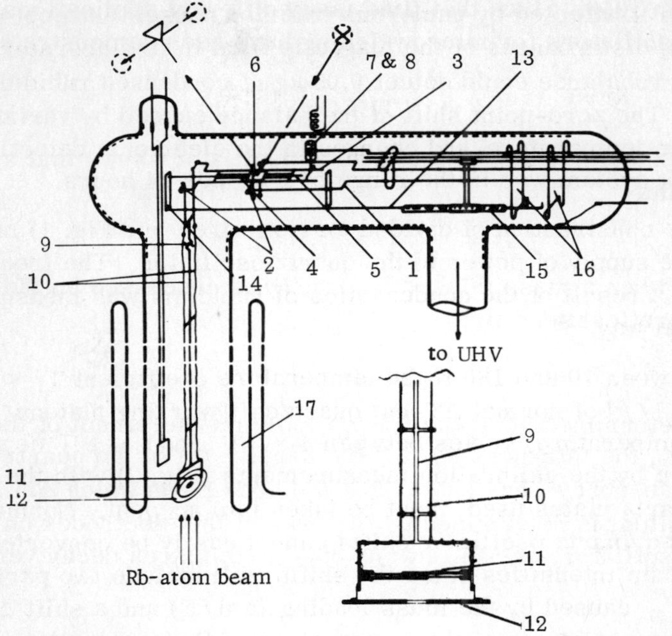

Fig. 1. Experimental arrangement for the comparison between the microbalance and the quartz oscillator. 1) Balance holder, 2) torsion-wire edge, 3) quartz scalebeam, 4) guiding tube, 5) quartz calibration wire, 6) mirror, 7, 8) permenant magnet surrounded by copper tube, 9) pan suspension, 10) silver wires, 11) quartz oscillator, 12) pan, 13) draw-beam, 14) cog, 15) guiding tube (attachment schematic), 16) bimetals, and 17) cooling liquid.

($\Delta m/F$) and the same atom-beam intensities. To minimize the effects of residual gases all measurements were made at pressures below 10^{-9} torr.

The mass condensed on the microbalance system is electromagnetically compensated by means of a feedback system,[15] which supplies the current for the compensation coil mounted outside the vacuum chamber, just over the permanent magnet that is fixed to one arm of the balance. The balance is calibrated by the shift of a homogeneous thin quartz fiber guided through a tube fixed to the quartz scalebeam. This in situ calibration, introduced first by Schmider,[16] permits calibration weights as small as 2 μg, and calibration during the measurements. This has proved to be extremely important, as the sensitivity of magnetically compensated balances is affected by numerous exterior causes (bakeout procedures, disturbances of the magnetic field in the laboratory, etc.). The microbalance could detect 0.05 μg of condensed rubidium very easily. The zero-point shift of the balance caused by variations of the room temperature and changes in the electronic detection and feedback system was in the range of 1 μg per 24 hours.

The construction of divided torsion wires (cf. Fig. 1) made possible the supply of power to the quartz oscillator. The frequency shift as a result of the condensation of rubidium was measured in the usual way.[11]

Between 70 and 180 K the temperature coefficient $T_f = [(df/dT)/f]$ of normal AT-cut quartzes ("working plateau" at room temperature) varies between 4×10^{-6} and 8×10^{-6} deg^{-1} and, as shown by the calibration measurements (Fig. 2) with the 5.5-MHz quartz plates used, must be taken into account. Normally the measured quartz oscillator shifts cannot easily be converted into atom-beam intensities since the shifts will include two parts: a shift Δf_m caused by the mass loading ($\Delta m/F$) and a shift Δf_T caused by variations of the temperature of the quartz plate. The expression Δf_T includes a time dependent $\Delta f_{T,1}$ as a result of the condensation of atoms on the quartz crystal, and a $\Delta f_{T,2}$ from the temperature changes of the surrounding cooling volume. For small deviations of the quartz oscillator temperature from the low equilibrium temperature it can be expected that the frequency shift, depending on condensation time t, is given by

$$\Delta f_T(t) \approx \Delta f_{T,1}(t) + \Delta f_{T,2}(t). \tag{3}$$

Theoretically, $\Delta f_{T,1}(t)$ is expected to be an exponential function of the form $1 - \exp(-t/\tau)$, so that its time dependence could be neglected for condensation times t much greater than the cooling period τ of the cooling ensemble. In this case $\Delta f_{T,1}(t)$ describes more or less the frequency change for a temperature increase, from the equilibrium temperature before condensation to the equilibrium temperature at a constant condensation rate.

The total temperature dependence of the quartz frequency, $\Delta f_T(t)$, could not be determined experimentally. However, it was possible to get $\Delta f_{T,2}(t)$ by measuring the frequency increase with time for a series of equivalent calibration runs without condensa-

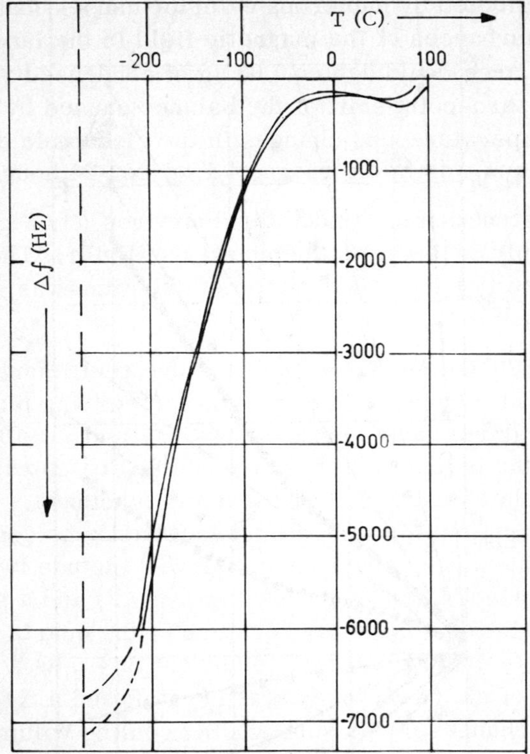

Fig. 2. Frequency shift Δf of a 5.5-MHz AT-cut quartz oscillator — "working plateau" at room temperature — versus temperature T in the range between 1.6 and 370 K.

tion. This could be done because a rough estimate showed that the cooling periods with or without condensation of rubidium did not differ considerably.

In Fig. 3 the deposited mass thickness as a function of condensation time t, as measured by the microbalance, is indicated by straight lines for four different Rb atom beam intensities between 10^{13} and 3×10^{14} atoms · cm^{-2} · sec^{-1} (curves 1a-4a). Schmider[16] and Göhre[17] have shown that this linearity is a criterion for 100% condensation. The plotted points (1b-4b in Fig. 3) correspond with the simultaneously measured quartz oscillator data, which do not agree with the microbalance data. If the quartz data are corrected

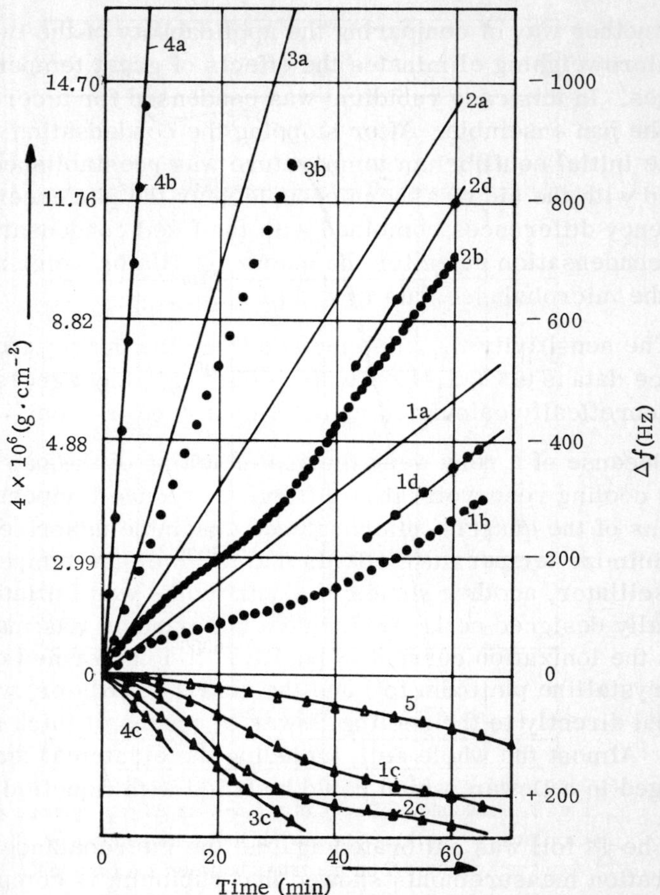

Fig. 3. Mass thickness φ versus condensation time t. a) Microbalance data; b) quartz data; c) extrapolated temperature influence; d) corrected quartz data. Curve 5 represents the mean average of the frequency shift Δf versus time as a result of a temperature increase of the surrounding Dewar.

for the temperature influence $\Delta f_{T,2}(t)$ — curve 5 shows the average of all calibration runs — one gets linear curves (e.g., curve 1d and curve 2d) parallel to the microbalance curves for $t \geq 40$ min. The determination of the condensation rates with the microbalance and with the corrected quartz data for ten different runs with varying atom-beam intensities agrees within ±2%. The curves with the index c are the extrapolated nonmeasured real temperature shifts Δf_T (t), which demonstrate the validity of Eq. (3).

Another way of comparing the applicability of the two systems for microweighing eliminates the effects of great temperature changes. In this case rubidium was condensed for a certain time onto the pan ensemble. After stopping the condensation by a shutter the initial equilibrium temperature was reestablished and controlled with the aid of a thermocouple near the pan ensemble. The frequency differences combined with the fixed condensation times gave condensation rates for the quartz oscillator which agreed with the microbalance data to within ±2.5%.

The sensitivity C_{exp} determined from the quartz and microbalance data is $(68 \pm 1.4) \times 10^6$ Hz · cm^2 · g^{-1} and agrees well with the theoretically calculated value of 68.3×10^6 Hz · cm^2 · g^{-1}.

Because of a very weak thermal coupling of the pan ensemble to the cooling reservoir, it is difficult to reduce temperature fluctuations of the quartz — microbalance ensemble described above. To minimize temperature effects and to lower the temperatures of the oscillator, another series of experiments was initiated in a specially designed cell. This time a comparison was made between the ionization current of positive Rb ions formed on a hot polycrystalline platinum foil and the quartz oscillator, which was coupled directly to the cooling Dewar by means of thick silver contacts. Almost the whole cell, including the electrical leads, was emerged in a Dewar, which could be filled with liquefied gas.

The Pt foil was calibrated against the microbalance.[18] These calibration measurements showed that rubidium is completely ionized on the hot polycrystalline platinum surface, the ionization coefficient being 1.0 between 1200 and 1900 K, which is in agreement with the Saha — Langmuir equation.

The temperature of the quartz oscillator was measured with a copper — constantan thermocouple, which was fixed outside the electrode area by means of a silver cement spot.

A typical run at a quartz-oscillator temperature of 81 K is shown in Fig. 4. The straight line indicates the frequency shift according to the measured Rb-ion current of 51 ± 1 Hz · min^{-1}. The dots represent the temperature-corrected quartz data according to a measured frequency shift of 50 ± 1 Hz · min^{-1}. The temperature influence in this experiment was low — a maximum rise of one degree was observed, which is equal to 45 Hz during the con-

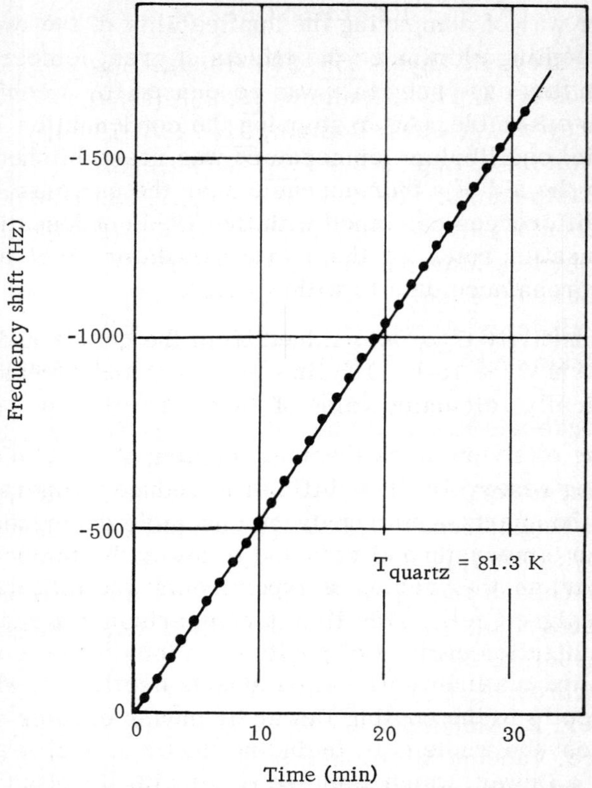

Fig. 4. Frequency shift versus condensation time; the straight line has been constructed with the measured Rb-ion current from a Pt surface (P. S. I.); the dots are the measured quartz data.

densation of Rb as a result of the good thermal coupling to the liquid-nitrogen bath used.

Both measurements — ion current and condensation rate — indicate the same Rb atom beam intensity of 10^{14} atoms $\cdot$ cm^{-2} $\cdot$ sec^{-1}. The sensitivity C_{exp} determined from these measurements is again $(68 \pm 1) \times 10^6$ Hz $\cdot$ cm^2 $\cdot$ g^{-1}, which is in excellent agreement with the other determinations.

Summing up all the results of these measurements, it can be stated that the quartz oscillator can be used without serious restrictions even at low temperatures. There is no indication of a change in sensitivity.

REFERENCES

1. G. Sauerbrey, Z. Phys., 155, 206 (1959).
2. M. P. Lostis, J. Phys. Radium, 20, 258 (1959).
3. J. C. Bruyère, J. Phys. Radium, 21, 222A (1960).
4. A. R. Wolter, J. Appl. Phys., 36, 2377 (1965).
5. T. E. Hartmann, J. Vac. Sci. Technol., 2, 239 (1965).
6. J. Edgecombe, J. Vac. Sci. Technol., 3, 28 (1966).
7. B. W. Kington, Marconi Instr., 10, 81 (1966).
8. W. H. Lawson, J. Sci. Instr., 44, 919 (1967).
9. H. K. Pulker, Z. Angew Phys., 20, 537 (1966).
10. H. J. Ishkin and V. S. Zazulin, Instr. Exptl. Techniques, 1, 44 (1963).
11. D. Hillecke and R. Niedermayer, Vakuum-Technik, 14, 69 (1965).
12. R. Niedermayer, N. Gladkich, and D. Hillecke, Vacuum Microbalance Techniques, Vol. 5 (K. H. Berndt, ed.) Plenum Press, New York (1966), p. 817.
13. H. L. Eschbach and E. W. Kruidhof, Vacuum Microbalance Techniques, Vol. 5 (K. H. Berndt, ed.), Plenum Press, New York (1966), p. 207.
14. G. Sauerbrey, A. E. Ü., 18, 617 (1964).
15. H. Mayer, R. Niedermayer, W. Schroen, D. Stünkel, and H. Göhre, Vacuum Microbalance Techniques, Vol. 3 (K. H. Berndt, ed.), Plenum Press, New York (1962), p. 75.
16. P. Schmider and H. Mayer, this volume, p. 207.
17. H. Göhre, Physik und Technik von Sorptions- und Desorptions vorgängen bei niederen Drucken, 2. Europäisches Symposium Vakuum, Frankfurt 1963.
18. D. Hillecke, Z. Phys., 215, 343 (1968).

Dynamic Vacuum in Microbalance Chambers

W. Kollen* and A. W. Czanderna

Department of Physics and Institute of Colloid
and Surface Science
Clarkson College of Technology
Potsdam, New York, 13676

ABSTRACT

The design of a microbalance vacuum system for obtaining a particular ultimate pressure in the sample region is examined in quantitative detail. The system considered includes the sample container, the hangdown tubes, balance housing, tubulation, and valves that lead to the pump. Since the ultimate pressure is simply the steady state between the gas influx into the system and the pumping speed, the basic equations for calculating the pressure and the effective pumping speed are considered. Then, the sources of total gas influx, such as outgassing of the chamber walls, permeation of gases through the walls, sample outgassing, desorption or outgassing, leaks, and any related temperature effects are considered for baked and unbaked Pyrex glass, quartz, and stainless steel systems. The equations for calculating the effective pumping speed at any point in a vacuum system with different sizes of tubulation and valves are considered. A working system is described to illustrate how the design considerations can be used to obtain the desired performance in a microbalance vacuum system. Finally, the ultimate pressure calculated for various gas loads is compared with the actual performance of the working system.

* Present address: Owens-Illinois, Inc., P. O. Box 1035, Toledo, Ohio.

INTRODUCTION

A review of the literature on microbalances shows that considerable time and effort have been devoted to the theory and design of the microbalance[1] but that little explicit information is available on the design of a microbalance vacuum system.[2] The importance of obtaining suitable vacuum has been generally recognized by a number of authors in the first few volumes in this series.[3–8] However, the design of a specific microbalance vacuum system by use of equations provided in any standard treatise on vacuum technique[9–11] and evaluation of the system by a performance study has not been reported.

In this paper, the dynamic vacuum in microbalance systems composed of a pump, valves, a balance housing with hangdown tubes, and the sample container will be considered. The ultimate vacuum at any point in a system depends on the steady state between the pumping speed and the gas influx. The pumping speed at any point in a vacuum system is determined by the size chosen for the pump, valves, and tubulation; it will remain constant after construction of a given system is completed. The total gas influx, which results from the net mass lost by the sample, leaks into the system, outgassing of the chamber walls, and the permeation of the gases through the walls, can vary by several orders of magnitude and thus change the ultimate pressure obtainable. The particular emphasis of this explicit development will be to design a system to provide a prechosen pressure around or inside a sample container that is suspended from the beam of a microbalance. This presumes that the experimenter will know what vacuum is necessary for successful study of a given sample. A secondary development will be to show that the system will have sufficient pumping speed when the gas influx increases rapidly during study, as might occur during decomposition or thermodesorption.

THEORY

The basic equation describing the pressure in a vacuum system is[11]

$$PS_e = -V(dP/dt) + Q \tag{1}$$

where P is the pressure, S_e is the effective pumping speed, V is the volume, and Q is the influx of gas into the system.* To obtain

* The units used in this paper for P, V, S_e, and Q are torr, liters, liters sec^{-1}, and torr · liter · sec^{-1}, respectively.

a low base pressure in a system, it is necessary to have a high effective pumping speed and a low influx. If a fast pumpdown is desired, a small volume is also necessary. It might appear from Eq. (1) that the volume of the chamber affects only the pumpdown speed but not the base pressure obtainable. However, since the surface area of the chamber usually increases with the volume, any additional influx resulting from the outgassing of the chamber walls will influence the base pressure. The base pressure, which is reached at the steady state between gas removal and gas influx, i.e., $dP/dt = 0$ in Eq. (1), is given by

$$P = Q/S_e \tag{2}$$

where P is now the base pressure. Most microbalance systems have relatively small volumes which allow the steady-state condition to be achieved rapidly after the start of pumping.

When a vacuum system is designed, it is likely that the pressure requirement in Eq. (2) will be selected first, depending on the intended application. It would be expected that, in general, the pressure requirement would be for the sample region. An examination of Eq. (2) indicates that the influx and the effective pumping speed may be varied to achieve the desired pressure. It will be shown below that the effective pumping speed depends on the size of the pump and the geometry of the path between the pump and the region of the vacuum system where a given pressure is desired. The influx in Eq. (2) may be derived from several sources. The gas may come from the outgassing of the chamber walls and tubulation, the permeation of gases through the walls, leaks, and from sample outgassing, desorption, or decomposition.

FACTORS AFFECTING THE INFLUX

All materials which have been exposed to a gaseous atmosphere will outgas when placed in a vacuum. Thus, the walls of the balance chamber, valves, and system tubulation, balance components, sample, etc., all become sources of gas and contribute to the influx. An extensive list of the outgassing properties for various materials at room temperature has been assembled by Dayton.[12] An extensive study of the outgassing of glass at various temperatures has been published by Todd.[13] The rate of outgassing will decrease with time on pumping at room temperature sometimes by as much as an order of magnitude in 10 h.[12] This is expected, since the outgassing rate is a desorption phenomenon. The net influx is related to the surface area, the surface coverage, the rate of adsorption, and the rate of desorption. At a given surface coverage, the outgassing rate should always increase with temperature.[14] In addition, diffusion of absorbed gases to the surface may become significant at higher temperatures. Thus, it is advisable to minimize the influx by maintaining the chamber first at a high temperature and then at as low a temperature as practical. Generally, this low temperature is room temperature for most applications.

Permeation is the diffusion of gases through the chamber walls into the system. For metals, hydrogen is the main component of permeation; for glass, both helium and hydrogen permeate with the rate for helium much greater. Permeation of gases is considered by many to be one of the limiting factors of the ultimate pressure in a glass system. It may not be a limiting factor in a Pyrex system with pressures above 5×10^{-12} torr if the system is continually pumped.[15] Specific influxes resulting from gas permeation from the atmosphere and calculated from available data[16,17] are listed in Table I. Partial pressures of helium and hydrogen in the atmosphere used in the calculations are 4×10^{-3} and 3.8×10^{-4} torr, respectively.[18] It can be seen in Table I that permeation is not likely to be a significant contributor to the total influx unless ultrahigh vacuum is desired and/or the system is operated at elevated temperatures.

Leaks and holes also contribute to the gas influx. With helium mass spectrometer leak detectors available at the present, the influx resulting from leaks can be reduced to less than 7.6×10^{-11} torr $\cdot$ liter $\cdot$ sec^{-1}.* Only in ultrahigh vacuum of approximately 10^{-11} torr or less would leak rates of this magnitude become significant.

A sample placed in the vacuum system will outgas. The influx in torr $\cdot$ liter $\cdot$ sec^{-1} from this can be calculated from the expression

$$Q = (62.36T/M) \cdot dm/dt \tag{3}$$

*For example, Veeco MS-9 or the equivalent are in widespread use; better instruments are available that will lower this figure by an order of magnitude.

Table I. Influx from Permeation of Various Materials

Material	Gas	Temperature	Specific influx,* $\frac{\text{torr} \cdot \text{liter}}{\text{sec} \cdot \text{cm}^2}$
Pyrex	Helium	25 C	2.66×10^{-15}
		400 C	3.74×10^{-12}
SiO_2	Helium	25 C	3.32×10^{-14}
		400 C	7.5×10^{-12}
Iron	Hydrogen	25 C	5×10^{-12}
		400 C	4×10^{-8}
27% Chromium steel	Hydrogen	25 C	1.3×10^{-15}
		400 C	1×10^{-9}

* Materials 1 mm thick.

where T is the temperature of the desorbing gas in °K, M is the gram molecular weight, and dm/dt is the rate of mass loss in $g \cdot sec^{-1}$.[19] The rate of sample outgassing is dependent upon its surface area, the activation energy of desorption of the adsorbed material, the frequency factor, and the surface coverage.[14, 33] In thermodesorption studies, the influx can be controlled by the rate at which the temperature of the sample is raised.

The two most common materials used for vacuum systems are glass and stainless steel. The choice usually depends upon the application. Some of the advantages of glass include ease of construction and repair, transparency, chemical inertness, low vapor pressure, and good electrical insulation properties. The main disadvantage is its poor resistance to mechanical shock. Stainless steel is strong, inert to most corrosive atmospheres, has a low thermal conductivity, and can be made in nonmagnetic forms; it is easily machined and can be welded with no fluxes by heli-arc welding. It can be baked to a higher temperature than glass. Stainless steel systems are not transparent and cannot be fabricated and altered as inexpensively as glass.

The size of the chamber should be large enough to provide sufficient pumping speed for the sample region. The shape should be cylindrical to keep the surface area as low as possible. In addition, a small chamber volume reduces the quantity of gas required to provide a desired pressure for studies in various gaseous atmospheres.

Frequently, values quoted for ultimate pressures in a system are meaningless unless they also correspond to the pressure in the sample region while data is being taken. The pressure in a vacuum system is not the same at all points in the system, especially when there is a significant amount of influx. For an accurate pressure measurement, the gauge should be located as near as possible to the region under consideration. Then, the measured pressure should be reported for the sample region. A fairly good estimate of the pressure in the sample region can be made if the gauge is located on the lower pressure end of the tubulation and if the size of the tubulation between the gauge and the sample region and the amount of gas given off by the sample are known.

The influx of gas from the walls of an unbaked chamber remains large enough, even on extended pumping, that it is seldom possible to attain pressures much below 10^{-7} torr with pumping speeds of less than 100 liter $\cdot$ sec^{-1}. To decrease the base pressure below 10^{-7} torr, then, either the effective pumping speed has to be increased or the rate of outgassing has to be reduced. Baking at 100C does not reduce the ultimate pressure obtainable in a system by a large factor.[20] A system has to be baked at temperatures much greater than 100C, preferably to 400C, for several hours while pumping to reduce significantly the ultimate pressure. The effect of baking Pyrex and stainless steel on the influx is listed in Table II.[21,22] From examination of Table II, it is seen that baking results in a decrease of influx of several orders of magnitude. This will permit the ultimate pressure of a system to be lowered considerably without employing expensive pumps. While the cost of the pump will be reduced substantially, there will be an additional cost to provide for baking.

Before a system is baked the primary residual gas is water vapor with small amounts of various other gases present. After baking and outgassing, carbon monoxide is the predominant residual gas in a glass system pumped by a mercury diffusion pump; hydrogen is the predominant residual gas in a stainless steel system pumped by an ion pump.[23]

Table II. Effect of Baking on the Influx from Pyrex and Stainless Steel at 25C

Material	Specific influx*	
	unbaked system	baked system
Pyrex glass	10^{-8}	10^{-15}
Stainless steel	2×10^{-8}	10^{-15}

* Influx in torr · liter · sec^{-1} · cm^{-2}.

When pumping with an oil diffusion pump it is nearly impossible to keep backstreaming of oil from entering the vacuum chamber.[24] This not only contaminates the system but also contributes to the influx of the system with a resultant higher ultimate pressure. Even with extensive trap systems it may not be possible to eliminate this backstreaming of oil vapor.[25] Thus, to ensure that the system is very clean, it should be baked and pumped with an ion pump; an extensive trap system must be employed if an oil diffusion pump is used.[24]

THE EFFECTIVE PUMPING SPEED

The effective pumping speed at the chamber can be calculated from the following expression[26]

$$\frac{1}{S_e} = \frac{1}{C} + \frac{1}{S_p} \tag{4}$$

where S_e is the effective pumping speed, S_p is the pumping speed of the pump, and C is the conductance between the pump and the chamber.

If Eq. (4) is inserted in Eq. (2),

$$P = \frac{Q}{C} + \frac{Q}{S_p} = P_c + P_p \tag{5}$$

where $P_c = Q/C$, the pressure drop across the conductance, and $P_p = Q/S_p$, the pressure at the pump. It is evident from Eq. (4)

that if the pumping speed is several times greater than the conductance, the effective pumping speed depends primarily on the conductance.

The value of the conductance will depend on the various items in series such as valves, tubing, and apertures. In this case the total conductance may be calculated from the expression

$$\frac{1}{C} = \frac{1}{C_1} + \frac{1}{C_2} + \frac{1}{C_3} + \ldots \tag{6}$$

where C is the total conductance and C_1, C_2, and C_3 are the conductances of the components of the system in series.

The conductance of a circular aperture in the pressure region of molecular flow is expressed as[28]

$$C_{SB} = 9.18D^2, \tag{7}$$

where C_{SB} is the conductance in liter · sec^{-1} for air at 20C, and D is the diameter in cm. The molecular flow pressure region is that region where the mean free path is much larger than the diameter of the tubing. For most microbalance systems pressures of less than 10^{-3} torr are in the molecular flow region.

It is important to have sufficient pumping speed at the sample to maintain low pressures. For example, during rapid thermodesorption and decomposition studies the pressure rise could result in force contributions from TMF effects if the pumping speed is very slow. The aperture of the sample holder bulb may seriously affect the pumping speed at the sample and, therefore, Eq. (7) is valuable for use when sample containers with small holes are employed.

If there is negligible outgassing of the tubulation from the pump to the chamber the conductance of a circular tube can be calculated in the pressure region of molecular flow using the relation[29]*

$$C_T = 12.12D^3/(L + 4D/3) \tag{8}$$

* When the source of gas is within the tubulation, the conductance should be calculated from the expression $C_T = 12.12D^3/L$.

where C_T is the conductance in liter · sec^{-1} for air at 20C, L is the length in cm, and D is the diameter of the tube in cm.

If outgassing of the tubulation is not negligible, Eq. (9) must be utilized to determine the pressure drop across a tube[30]:

$$\Delta P = qL^2/24.24D^2 \tag{9}$$

where ΔP is the pressure drop in torr for air at 20 C resulting from outgassing of the tubulation, L is the length of the tube in cm, q is the specific influx in torr · liter · sec^{-1} · cm^{-2} of the material, and D is the diameter of the tube in cm.

The total pressure drop P'_c, across a tube connecting a chamber and a pump becomes

$$P'_c = (Q/C) + P = P_c + \Delta P \tag{10}$$

and the pressure in the chamber is given by

$$P = P_c + \Delta P + P_p \tag{11}$$

The differential pressure P_c between the sample and any other region of the system resulting from the desorption of gases by the sample can be calculated by Eq. (12) if the rate of mass loss and conductance are known:

$$P_c = Q/C \tag{12}$$

where P_c was defined in Eq. (5), Q in Eq. (3), and C in Eq. (6). Thus, if the allowable pressure rise is known, the maximum rate of sample mass loss can be calculated for a given system by means of Eqs. (3), (5), (6), and (12). The special merit of using a microbalance is that dm/dt in Eq. (3) is measured directly.

DESIGN OF A SPECIFIC SYSTEM

The design of the vacuum system described below required the use of a Cahn RG microbalance for the measurement of the thermodesorption of oxygen from silver powder. It is not possible to bake this system because the Cahn balance will not withstand temperatures above 100 C. A base pressure requirement of 10^{-6} torr was chosen for the sample region. During thermodesorption, the maximum influx from the sample was estimated to be 7×10^{-3} μg · sec^{-1} or 4×10^{-6} torr · liter · sec^{-1} at 25 C. The use of a 1.25-in.-bore tube furnace for heating the sample to 400 C required the hangdown tubes to be 25 mm or less in diameter. The hangdown tubes were made as long as space and conductance requirements allowed because the operating performance of any microbalance is improved when the sample is suspended from a long fiber. Finally, it was necessary to insert a valve between the pump and the balance chamber to allow use of the system at high pressures. The final design of the vacuum system after imposing the above restraints is sketched in Fig. 1.

The balance chamber was connected to a 50 liter · sec^{-1} ion pump[31] with 3 in. diameter Pyrex tubing and a 3 in. stainless steel bellows type high-vacuum valve.[32] These sizes were selected from equipment available commercially after calculating the conductance necessary to achieve the chosen pressure of 10^{-6} torr. The ion pump was selected to eliminate the need for an extensive trapping system for the oil vapor. The Pyrex glass balance housing was joined to the main pumping tube at a ground glass joint sealed with Apiezon W wax;* this joint replaced the aluminum plate and O-ring available from the manufacturer of the balance. Electrical feedthroughs for the balance were made from capillary tubing and sealed with Apiezon W. Stainless steel flanges sealed with gold O-rings were provided in the hangdown tubes for access to the samples. A cylindrical container 1 cm in diameter and 1.5 cm long with the top completely open was used to suspend the silver powder from the balance. This design was necessary to provide an adequate conductance at the sample container relative to the conductance of the hangdown tube.

The influx of gas in this system results primarily from outgassing of vacuum system components and the desorption of gases from the sample. The permeation of Pyrex glass at pressures of 10^{-6} torr is negligible at room temperature. (Compare data in Tables I and II.)

The pressure in the system was measured using a Veeco RG-75P hot cathode ionization gauge. The gauge was positioned on a

* Apeizon W wax may be used quite safely at pressures as low as 10^{-8} torr.

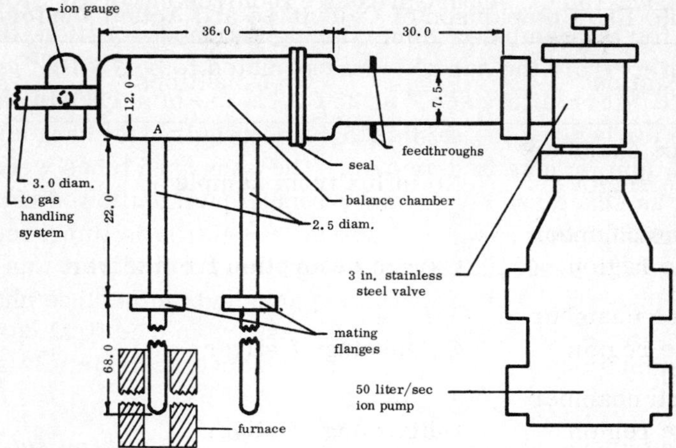

Fig. 1. Vacuum system for microbalance; dimensions in cm, except for valve.

connecting tube to the balance chamber near the sample hangdown tube. Calculations of the conductance show that a pressure measured at this gauge position corresponds very closely to the pressure at the opening of the sample hangdown tube (A, Fig. 1). As indicated in the preceding sections, the pressure in the sample region can be estimated quite well from the pressure reading at A.

A stainless steel bellows type high-vacuum valve was used to connect a gas-handling system to the vacuum chamber. The roughing system was incorporated into the gas handling system. It consists of a 25 liter sec^{-1} CVC oil diffusion pump with a Linde 13X molecular sieve trap backed by a mechanical pump. The roughing system was used to pump the balance chamber down to the 10^{-4} torr range before switching to the ion pump. This roughing with a mechanical pump and then a slow-speed diffusion pump permits the ion pump to be started without heavy discharge. If the flow-through in pumping with an oil diffusion pump is quite large, there is very little backstreaming of oil vapors. Thus, an oil pump may be used for roughing with negligible oil contamination.

RESULTS

The calculated and actual performance of the specific microbalance vacuum system is summarized in Table III. The calcu-

Table III. Comparison of Calculated and Actual Performance of

Location	Conditions*
Balance chamber	
Sample region	No influx from sample
Balance chamber	
Sample region	Oxygen desorption from silver
Balance chamber	
Sample region	Outgassing of silver
Balance chamber	
Sample region	Outgassing of silver
Balance chamber	
Sample region	Outgassing of silver
Balance chamber	
Sample region	Slow $O_2\uparrow$ by decomp. of $KMnO_4$ at 250 C
Balance chamber	
Sample region	Decomp. of 0.65 g NaN_3 at 300 C in 8 h
Balance chamber	
Sample region	Rapid dehydration of 0.227 g $NiCO_3 \cdot 6H_2O$ at 25 C in 1000 sec.

*Stated only for the sample. The influx from outgassing of the Pyrex walls in the entire system is assumed to remain constant at 10^{-8} torr · liter · sec^{-1} · cm^{-2}

[a] The pressure difference between the balance chamber and the sample region results from considering the conductance of the influx from the hangdown tube.

[b] Estimated by using a value for q that produces a calculated chamber pressure of 6.9 × 10^{-7} torr with no influx from sample.

[c] Estimated using q as note (a) and influx from sample.

lated pressures were obtained by using the expressions and data presented in the previous section. The influx resulting from the balance weighing mechanism was estimated to be 1.25×10^{-5} torr liter · sec^{-1} and that of the valve 5×10^{-6} torr · liter · sec^{-1}. It may be seen in Table III that pressures in the 10^{-6} torr range are indeed possible in the sample region with an influx from the sample up to 4×10^{-6} torr · liter · sec^{-1}. In addition, the pressure measured at the balance chamber is within a factor of three of the

Microbalance Vacuum System at a Varying Influx from the Sample

Influx from the sample, torr · liter · sec^{-1}	Calc. pressure, torr	Measured pressure, torr
0	1.58×10^{-6}[a] 3.21×10^{-6}[a]	6.9×10^{-7} 1.5×10^{-6}[b]
4×10^{-6}	1.68×10^{-6} 5.65×10^{-6}	9.5×10^{-7} 3.05×10^{-6}[c]
2.54×10^{-5}	2.36×10^{-6} 1.88×10^{-5}	2.54×10^{-6} 1.65×10^{-5}[c]
3.52×10^{-5}	2.68×10^{-6} 2.49×10^{-5}	3.52×10^{-6} 2.48×10^{-5}[c]
6.5×10^{-5}	3.63×10^{-6} 4.32×10^{-5}	6.5×10^{-6} 4.52×10^{-5}[c]
1.0×10^{-4}	4.74×10^{-6} 6.48×10^{-5}	
1.0×10^{-2}	3.17×10^{-4} 6.17×10^{-3}	
1.12	3.56×10^{-2} 6.91×10^{-1}	

actual pressure in the sample region. However, when the sample outgassing is in the 10^{-5} torr · liter · sec^{-1} range, the measured pressures may be at least an order of magnitude too low. During thermodesorption of oxygen from silver powder, the maximum influx observed is 4×10^{-6} torr · liter · sec^{-1}. This results in an increase of the pressure in the sample region to about 3.0×10^{-6} torr. Thus, pressures in the lower 10^{-6} torr range while data of thermodesorption are taken have been achieved with this system.

CONCLUSIONS

The results show that by quantitative consideration of the influx and conductance, a microbalance vacuum system can be designed to achieve a chosen pressure with a specified gas load. Pressure in a system varies considerably depending on a number of conditions. Thus, pressure measurements are of little value unless they adequately represent the pressure in the region questioned while data are taken.

ACKNOWLEDGMENTS

The authors are pleased to acknowledge partial support of this project by the Division of Research at Clarkson College of Technology. Acknowledgment is made to the donors of the Petroleum Research Fund, administered by The American Chemical Society, for partial support of the project. Partial support for this work by the Division of Air Pollution, Bureau of State Services, Public Health Service under Research Grant No. 1RO1 AP00552 01, is gratefully acknowledged.

REFERENCES

1. R. L. Schwoebel, Microbalance theory and design, in: Ultra Micro Weight Determination in Controlled Environments (S. P. Wolsky and E. J. Zdanuk, eds.), Interscience-Wiley, New York, (1969), p. 61.
2. A. W. Czanderna, Ultramicrobalance review, Ibid, p. 7.
3. E. A. Gulbransen and K. F. Andrew, An enclosed physical chemistry laboratory, in: Vacuum Microbalance Techniques, Vol. 1 (M. J. Katz, ed.), Plenum Press, New York (1961), p. 10.
4. S. P. Wolsky and E. J. Zdanuk, Ibid., p. 35.
5. K. F. Behrndt, Ibid., p. 69.
6. G. F. Rouse, Vacuum Microbalance Techniques, Vol. 2 (R. F. Walker, ed.), Plenum Press, New York (1962), p. 59.
7. A. W. Czanderna and H. Wieder, Ibid., p. 147.
8. N. J. Carrera et al., Vacuum Microbalance Techniques, Vol. 3 (K. H. Berndt, ed.), Plenum Press, New York (1963), p. 153.
9. S. Dushman and J. M. Lafferty, eds., Scientific Foundations of Vacuum Technique, 2nd Edition, John Wiley & Sons, New York (1962).

10. R. W. Roberts and T. A. Vanderslice, Ultra High Vacuum and Its Applications, Prentice Hall, Englewood Cliffs, New Jersey (1963).
11. C. M. Van Atta, Vacuum Science and Engineering, McGraw-Hill, New York (1965), p. 274.
12. B. B. Dayton, Relations between size of vacuum chamber, outgassing rate, and required pumping speed, in: 1959 6th National Symposium on Vacuum Technology Transactions (C. R. Meisser, ed.), Pergamon Press, New York (1960), p. 101.
13. B. J. Todd, Outgassing of glass, J. Appl. Phys., 26, 1238 (1955).
14. D. O. Hayward and B. M. W. Trapnell, Chemisorption, Butterworths, London (1964).
15. B. J. Hopkins and K. R. Pender, A liquid helium cooled trap for ultrahigh vacuum systems, J. Sci. Inst., 44, 73 (1967).
16. R. W. Roberts and T. A. Vanderslice, op. cit., pp. 100, 110.
17. F. J. Norton, J. Am. Ceram. Soc., 36, 90 (1953); Gas permeation through the vacuum envelope, in: 1961 Transactions of the Eighth National Vacuum Symposium, Vol. 1 (L. E. Preuss, ed.), Pergamon Press, New York (1962), p. 8.
18. Compendium of Meteorology, American Meteorology, American Meteorol. Soc., Boston (1951), p. 6.
19. C. M. Van Atta, op. cit., p. 24.
20. Ibid., p. 364.
21. A. H. Turnbull, R. S. Barton, and J. C. Riviere, An Introduction to Vacuum Technique, John Wiley & Sons, New York (1963), p. 167.
22. B. B. Dayton, op. cit., p. 101.
23. B. J. Hopkins, Ultrahigh vacua, Contemp. Phys., 9, 115 (1968).
24. R. D. Gretz, J. Vac. Sci. Tech., 5, 49 (1968).
25. D. H. Buckley, M. Swikert, and R. L. Johnson, Friction, wear, and evaporation rates of various materials in vacuum to 10^{-7} mm Hg, ASLE/ASME Preprint No. 61 LC-2, Academic Press, New York (1961).
26. C. M. Van Atta, op. cit., p. 25.
27. Ibid., p. 25.
28. Ibid., p. 48.
29. Ibid., p. 50.
30. D. J. Santeler, Pressure measurement in nonbaked ultrahigh vacuum systems, in: 1961 Transactions of the Eighth National

Vacuum Symposium, Vol. 1 (L. E. Preuss, ed.), Pergamon Press, New York (1962), p. 549.

31. Data Sheet, Vacion Pumps, 50 liter/sec, Varian, Vacuum Division, Palo Alto, California 94303, VAC 2279, Oct. 1966.

32. Bulletin B-1359, Ultek Corp., Palo Alto, California 94303.

33. A. W. Czanderna, J. R. Biegen, and W. Kollen, J. Colloid Interface Sci., 34 (1970), in press.

Methods for the Elimination of Weighing Troubles Due to Convection in a Microbalance

W. Kuhn, E. Robens, G. Sandstede, and G. Walter

Battelle-Institut e.V.
Frankfurt/Main, W. Germany

ABSTRACT

The paper deals with an investigation into the fluctuations encountered in thermogravimetric measurements using a microbalance with suspended pans, in a temperature range up to 1200 C. The convection resulting at gas pressures above 200 torr, which is responsible for these fluctuations, depends on the pressure, the type of gas used, the temperature gradient, and the geometry of the arrangement. The difficulties have been overcome by using long, thin suspension wires; shielding samples and balance beam with horizontally suspended foils attached to the suspension wire or to the balance tube; using heaters with a homogeneous temperature zone and a helium gas stream protecting the measuring system.

INTRODUCTION

If sample and measuring system in a balance are at different temperatures, convective gas currents may be produced so that the moving parts of the balance are subjected to frictional forces. This leads to apparent weight changes and to fluctuations of the readings.

At pressures below, say, 1 torr, radiometer forces interfere with the measurements.*[1] In the range between 1 and 200 torr

* Literature references on the disturbances at low pressures are included in reference 2.

only minor disturbances have been observed. Above 200 torr, the readings show large fluctuations. In a balance with suspended pans the fluctuations are particularly large if the sample is at a higher temperature than the measuring system.*

The disturbances caused at pressures above, say, 1 torr are due to convection. Their influence is difficult to predict as the profile of convection currents depends not only on the gas pressure and the temperature difference between sample and measuring system, but also on the specific temperature distribution and the geometry of sample and apparatus. Several solutions which avoid these difficulties have been proposed for thermogravimetric instruments.[4–6] Our paper is concerned with investigations covering the main parameters of convection in microbalances with suspended pans, and includes various proposals for minimizing the reading errors.

EXPERIMENTAL INVESTIGATIONS

Investigations Using a Microbalance in an Aluminum Casing

For investigating catalytic reactions[7] we enclosed a Gast microbalance[8] in an aluminum casing (Fig. 1). The two balance tubes had to be comparatively short. They were made from an aluminum block and connected with each other through a bore at the level of the balance pans. However, even at a temperature difference of 100C between sample and measuring system and a pressure of 100 torr, this arrangement caused a thermosiphon current which led to a constant apparent change in weight of more than 1 mg. The aluminum tubes were then replaced by tubes of glass which were not connected with each other. By means of radiation shields fixed to the suspension wires above the pan and below the beam, it was possible to reduce the disturbances to a few micrograms at sample temperatures up to 300C (measuring system at room temperature). Major fluctuations of the readings only occurred at gas pressures above 500 torr in gases such as argon or nitrogen.

* The inverse case, sample at a lower temperature than the measuring system, is discussed in a separate paper.[3]

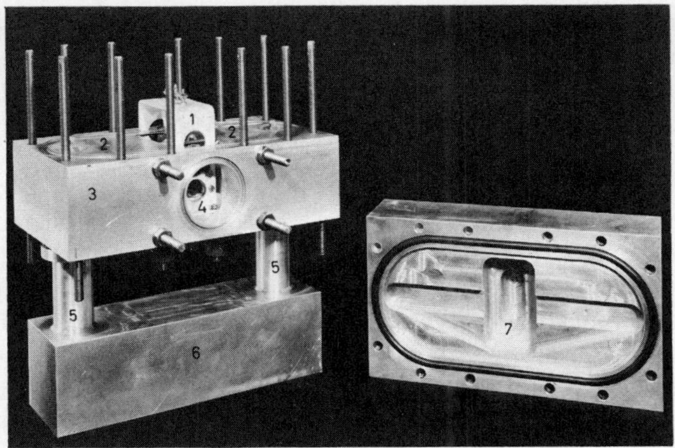

Fig. 1. Sartorius microbalance in aluminum casing. 1) Measuring system; 2) balance beam; 3) aluminum casing; 4) pump nozzle; 5) balance tubes; 6) connection tube; 7) cover of aluminum casing.

Investigations Conducted by Cahn and Schultz

Cahn and Schultz[4] report experiments made on differently shaped samples in the usual balance arrangement with relatively short balance tubes. In air at a pressure above 150 torr and a temperature of 100C (measuring system at room temperature) they observed irregular fluctuations of the readings, with a maximum amplitude of about 300 μg. The amplitudes increased slightly with rising temperature. Between 200 and 500 torr they increased strongly with increasing pressure and above 500 torr they remained constant. Only the horizontal faces of the sample contribute to the fluctuations, which are proportional to the area of the horizontal faces. In order to minimize this disturbance, the authors recommended the use of thin balance tubes (7 mm in diameter) provided with horizontal shields. They also recommended that the sample be enclosed in a metal cup which has a small opening in its lid for the suspension wire.

Investigations Using the Gravimat

In order to study measures to eliminate convective fluctuations in the Gravimat* system, we used the arrangement shown in Fig. 2.

* Manufacturer of the Gravimat[9] and of the Gast-type microbalance[8]: Sartorius-Werke GmbH, Göttingen, W. Germany.

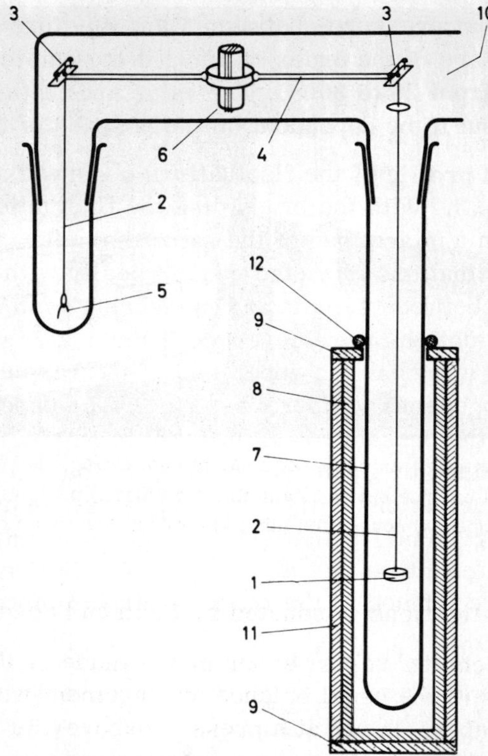

Fig. 2. Arrangement with Gast microbalance. 1) Brass disk, diameter 10 mm; 2) suspension wire, diameter 30 μ; 3) sapphire pans on diamond pins; 4) balance beam; 5) counterweight; 6) coil system and magnet; 7) balance tube wrapped in wire gauze for grounding; 8) aluminum tube wrapped in fiberglass; 9) asbestos cover; 10) nozzle to vacuum pump, gas supply, and pressure control; 11) heater; 12) asbestos cord.

The balance carried a brass disk (diameter 10 mm). The disk was suspended 850 mm below the beam in a balance tube which had an inside diameter of 25 mm. Wire gauze protected the balance tube from static charging. By means of a tube heater the temperature in the balance tube was kept constant with an accuracy of ± 1 C over a length of about 100 mm. The measuring system and the counterweight, a wire hook, were at room temperature in all experiments.

In nitrogen at pressures between 1 and 300 torr the apparent weight changes are of the order of 10 μg if the pan temperature was increased from 25 to 300 C. The value and the sign of this deviation turned out to be dependent on the position of the heater.

The typical profile of the fluctuations observed above 300 torr is shown in Fig. 3. With the brass disk the fluctuations were larger than with a quartz pan of the same diameter. The amplitudes of the fluctuations increase monotonically with the temperature difference between sample and measuring system. Particularly large fluctuations were observed if the heater was lowered so that the disk was near the upper rim of the heater and thus in a highly inhomogeneous temperature zone. The dependence of the fluctuations on the pressure is illustrated in Fig. 4. In contrast to the results obtained by Cahn and Schultz, we found a monotonic increase of the fluctuations with rising pressure. In hydrogen and helium we observed no fluctuations. From the results of these investigations we conclude that above a specific pressure convective vortices are formed in the range of large temperature inhomo-

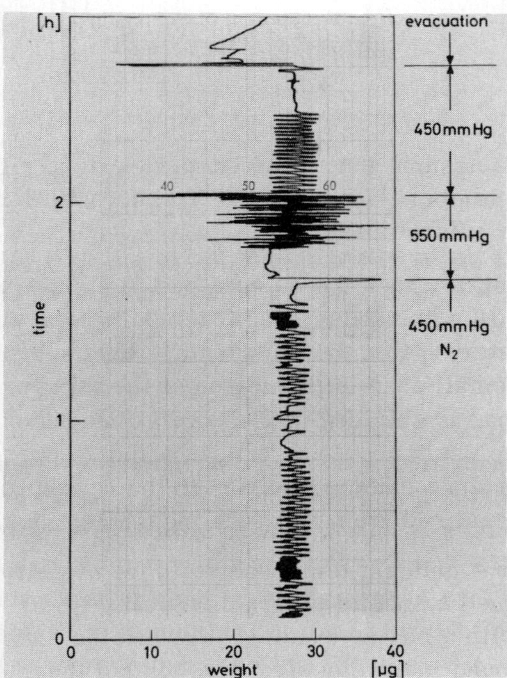

Fig. 3. Fluctuations of readings due to convection.

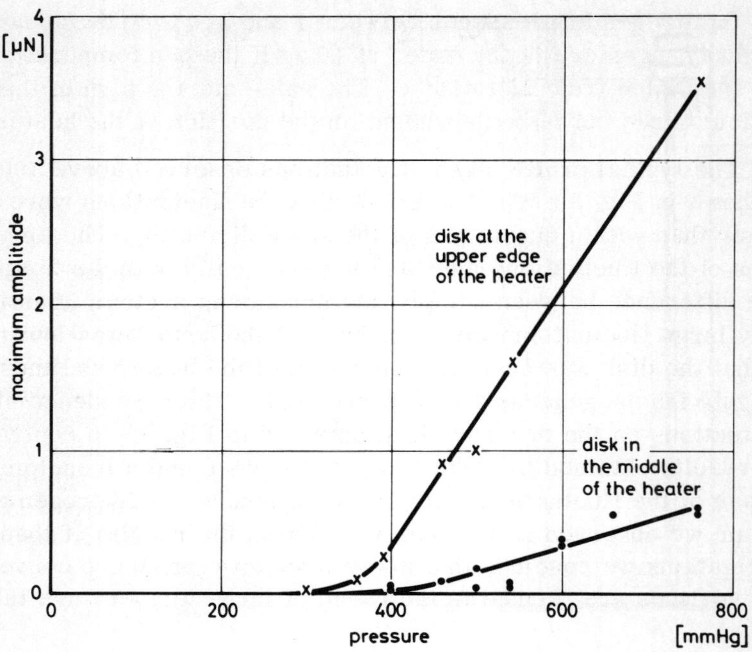

Fig. 4. Dependence of the fluctuations of the readings on gas pressure and position of the heater.

geneities. We assume that these vortices extend right into the homogeneous temperature zone and interact with large faces of the sample or the balance beam.

We therefore attach the balance pan to a suspension wire 1 m in length and 30 μ thickness, and at temperatures above 600 C we shield the heater so that the temperature inhomogeneities are restricted to a small zone in the region of the thin suspension wire. The balance pan is shielded with a metal disk attached to the suspension wire closely above the pan, in order to prevent the formation of temperature inhomogeneities in the region of the pan as a result of thermal radiation from the measuring system.[3]

To ensure a uniform temperature in the vicinity of the balance pan, we designed a special external heater (Fig. 5). It consists of a steel tube with a bifilar winding. Toward the ends of the tube the filament is wound more closely. The control thermocouple is in

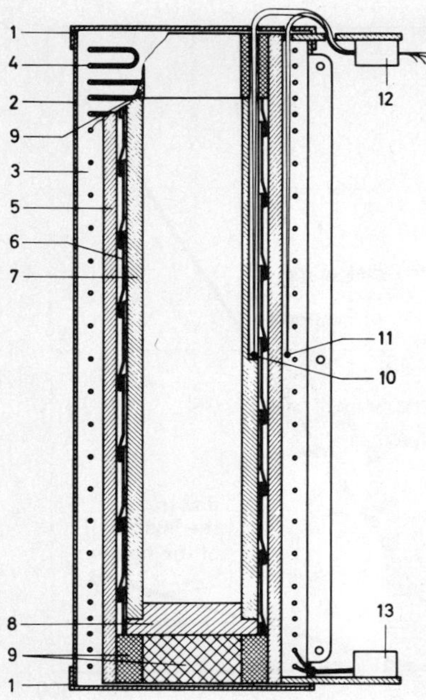

Fig. 5. Heater for vacuum microbalance. 1) Cover (2 mm, St. 0012); 2) sleeve (2 mm, St. 0012); 3) Fiberfrax; 4) bifilar winding; 5) outer tube (5 mm, St. 4841); 6) fiberglass; 7) inner tube (light metal); 8) plug (light metal); 9) asbestos board; 10) thermocouple for temperature measurement (Ni – CrNi); 11) thermocouple for control (Ni – CrNi); 12) connection for thermocouples; 13) terminal for filament winding.

close proximity to the filament winding. A light metal tube wrapped in fiberglass is placed loosely in the steel tube. This involves poor thermal conduction with respect to the filament, while the heat exchange along the tube is good. The wall of the light metal tube is provided with a groove which contains a thermocouple for temperature measurement. The heater has a detachable bottom so that it can be fitted over any part of a tube. In this design the heater is suited for temperatures up to about 600C. The temperature produced by the heater in the middle of the balance tube is kept constant with an accuracy of 1 C over a length of about 100 mm.

Disturbances in the temperature range up to 1200 C and in the pressure range up to atmospheric pressure were reduced by these measures to a few μg.

Investigations with Six Microbalances in a Common Vessel

Figure 6 shows an apparatus, with six microbalances in a common vessel, which was used for investigating the corrosion in a flowing gas in the presence of nuclear radiation.[10] Figure 7 shows the balance system (Sartorius) and its support. The design of the

Fig. 6. Apparatus for measurement of corrosion in the presence of nuclear radiation. 1) Vessel; 2) microbalances; 3) reaction tube; 4) samples; 5) β-source; 6) reaction gas inlet; 7) helium inlet; 8) gas outlet; 9) shields against thermal radiation; 10) turbomolecular pump; 11) rotary pump; 12) stainless steel shield against nuclear radiation; 13) suspension wire for lifting the β-source; 14) diffusion pump; 15) butterfly valve; 16) lead shield against nuclear radiation.

apparatus is illustrated in Fig. 8. With this apparatus we investigated the corrosion of sheets of iron, zirconium, and stainless steel by continuous weight determinations in flowing gases and gas mixtures (oxygen, hydrogen, water vapor). The temperature of the samples was 300 C, that of the measuring systems 120 C. Samples and measuring systems were partially shielded against thermal radiation by means of metal sheets permanently installed in the reac-

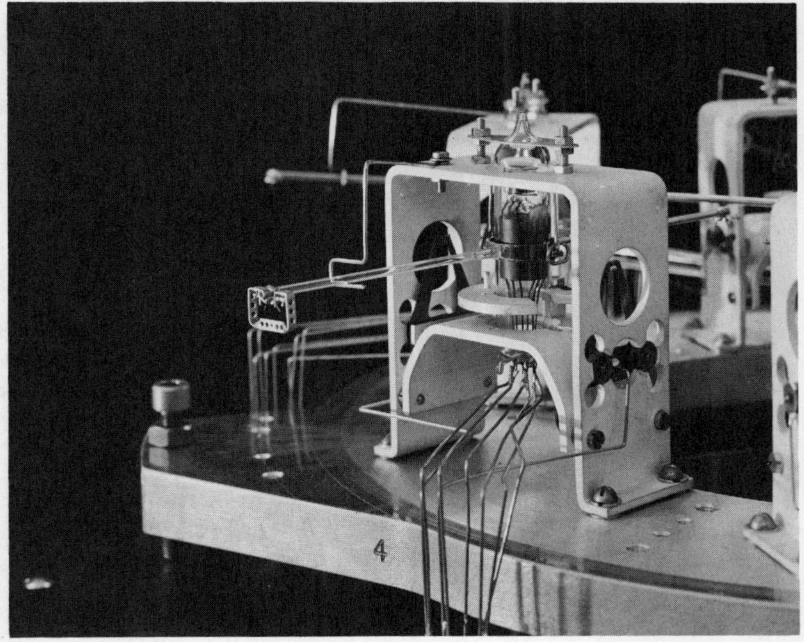

Fig. 7. Sartorius microbalances on aluminum support.

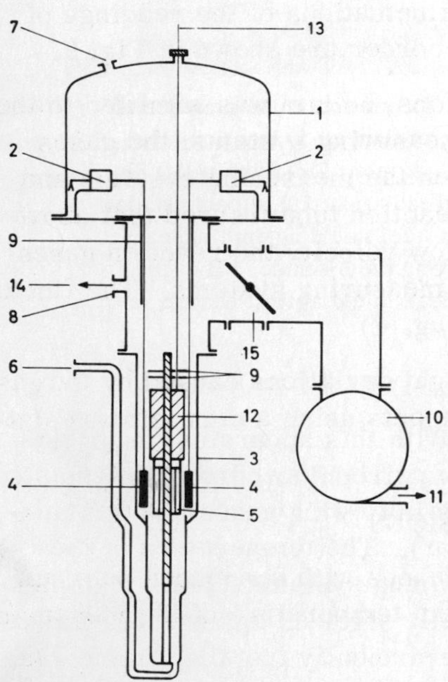

Fig. 8. Apparatus for measurement of corrosion in the presence of nuclear radiation, diagrammatic view (for denotations see Fig. 6).

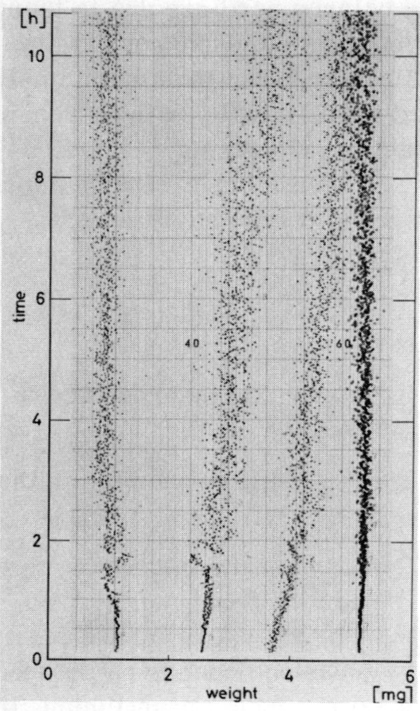

Fig. 9. Fluctuations of the readings of four microbalances in a common vessel due to convection; measuring system 120 C, samples 200 C, O_2 at 700 torr.

tion tube. Because of the rather unfavorable shape of the sample tube, which had to be used, fluctuations of the order of 1 mg occurred even at temperature differences of 100 C between samples and measuring systems. Typical fluctuations of the readings of four balances plotted by a point recorder are shown in Fig. 9.

In order to check the fluctuations, helium was admitted to the part of the vessel containing the measuring systems; the gases were drawn off in the space between the measuring systems and the samples. The shields in the reaction tube insured that above a specific empirically determined flow velocity the reaction gases did not come into contact with the measuring systems. The fluctuations thus were reduced to a few μg.

The apparent c o n s t a n t weight deviations caused by the gas flow were determined by measurements using a corrosion-resistant dummy.

CONCLUSIONS

In our investigations using balances with suspended pans and sample temperatures higher than the temperature of the measuring system irregular fluctuations of the balance readings due to thermal convection were observed in various gas atmospheres (argon, air, nitrogen), at pressures above 300 torr. The maximum amplitudes of these fluctuations increased monotonically with increasing sample temperature and with increasing gas pressure. Particularly large fluctuations were observed when a large-sized sample was suspended in a highly inhomogeneous temperature zone.

We avoided these difficulties by using

(1) long and thin suspension wires;

(2) shields attached to the suspension wire or incorporated in the balance tube;

(3) a homogeneous temperature zone as long as possible in the vicinity of the sample;

(4) a hydrogen or helium stream protecting the measuring system from the reaction gas.

The efficiency of the individual measures depends on the experimental conditions and on the geometry of the arrangement. In a later investigation, the effectiveness of these measures may be studied in more detail in order to develop simpler designs.

ACKNOWLEDGMENTS

For valuable contributions we are grateful to Dr. G. Wurzbacher; for the performance of the measurements our thanks are due to Mrs. B. Krafczyk and Mr. H. Huser, Battelle-Institut e.V., Frankfurt/Main.

REFERENCES

1. H. Krupp, G. Walter, E. Robens, and G. Sandstede, Vacuum, 13, 197 (1963).
2. E. Robens and G. Sandstede, Z. Instrumentenkunde, 75, 125 (1967).
3. E. Robens, G. Sandstede, G. Walter, and G. Wurzbacher, paper in this volume, 195-205.
4. L. Cahn and H. Schultz, Anal. Chem., 35, 1729 (1963).
5. C. Duval, Inorganic Thermogravimetric Analysis, 2nd Edition, Elsevier, Amsterdam (1963).

6. P. D. Garn, Thermoanalytical Methods of Investigation, Academic Press, New York (1965).
7. G. Walter and G. Wurzbacher, Sorption of gaseous hydrocarbons at fuel cell catalysts of the platinum metal group, Final Technical Report AD-467,849, Clearinghouse for Scientific and Technical Information, Springfield (1965).
8. Th. Gast, Vakuumtechnik, 14, 41 (1965).
9. E. Robens and G. Sandstede, Rev. Sci. Instr. Series 2, 2, 365 (1969).
10. W. Kuhn and G. Walter, Microgravimetric investigation into the mechanisms of corrosion of reactor materials in the presence of nuclear radiation, Euratom Report 1474. e. EURAEC Rep. No. 874, Presses Académiques Européennes, Brussels (1964).

Comments on the Applications and Improvement of a UHV Microbalance

H. Moret, E. Louwerix, and E. Sattler

Central Bureau for Nuclear Measurements
EURATOM
Geel, Belgium

ABSTRACT

The balance reported at the 5th Conference has been used in several studies (e.g., the determination of masses of layers which are evaporated in vacuum and ultrahigh vacuum, and the determination of drop weights). The results of these studies provided useful information on the best precision and accuracy which can be achieved under different environmental conditions. Tests of reproducibility and linearity of response showed that accuracies of $\pm 1\ \mu g$ and better can be realized, but only with very careful control of operating conditions, especially when mass differences exceeding 1 or 2 mg are to be measured. Several modifications of the electronics (increased gain of feedback amplifier) and the mechanical construction (more rigid suspension) improved the reliability and ease of operation of the balance.

INTRODUCTION

At a former Conference, a microbalance for ultrahigh vacuum applications was reported.[1] Its main characteristics (see Fig. 1) are:

useful load of 2.5 g;

reproducibility, 0.3 μg (standard deviation);

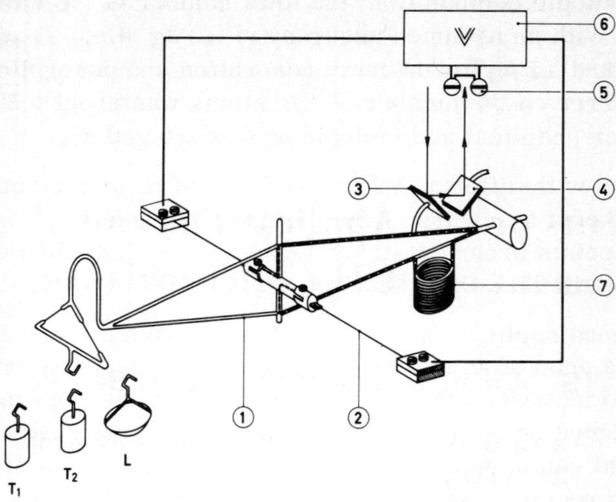

Fig. 1. UHV microbalance. 1) Beam; 2) main suspension; 3) deflection mirror; 4) reference mirror; 5) photoresistors; 6) amplifier; 7) coil.

bakeable up to 400°; vacua in the range of 10^{-10}–10^{-11} torr are readily obtained;

substitution weighing, thereby eliminating drift of zero and electrical sensitivity.

A number of these balances have been used for different applications. Two examples are shortly reviewed.

Experience under various conditions led to some minor modifications to the mechanical design. A complete redesign of the electronics and the electrical measuring circuit improved the dynamic performance and the reproducibility. Also, operation and reading of the balance became more convenient. Preliminary tests seem to indicate perfect linearity of response over the full electrical range from −10 mg to +10 mg.

APPLICATIONS

In one application, boron layers were prepared by vacuum evaporation. The mass of deposited material could be determined with the balance. After applying corrections for chemical impurities and isotopic composition, the total number of ^{10}B atoms was obtained. With an assumed accuracy of ± 2 μg (i.e., ± 1 μg weighing error and ± 1 μg for unknown adsorption and desorption effects) the total error on the number of ^{10}B atoms was about 0.5%, including 0.2% for chemical and isotopic analysis together.

Layers with different thicknesses could be intercompared in a neutron beam by means of the $^{10}B(n, \alpha)^7Li$ reaction.[2] Scattering of the points around a straight-line fit was less than 0.3%, which is better than was expected.

A second application concerns the determination of the initial weight of a drop of an aqueous solution which is deposited on a thin foil.[3] Two methods are currently used. With the first method the foil is weighed on a conventional balance; the first reading of the drop weight can normally be taken 1-2 min after deposition, this time being required by the balance to regain equilibrium. Plots of drop weight as a function of time indicate a constant evaporation rate after 1 min. Between zero time and 1 min, however, very significant deviations occur, which may lead to errors up to 50 μg when linear extrapolation is applied. More reliable extrapolation is obtained if quicker response of the balance can be realized. Therefore, the automatic balance was used for this method.

The second method is to weigh the dispensing pycnometer before and after expelling the drop. The rate of evaporation from the pycnometer is much smaller and is constant in time. In order to confirm the validity of both methods they were applied simultaneously with the same drop and with the same balance, i.e., the vacuum balance. A mechanical device allowed the foil and the pycnometer to be suspended alternatively from the balance. Mass variations in time for both methods were obtained. Extrapolations to zero time (time of deposition of the drop) showed that within experimental error no differences between the methods could be found.

LIMITATIONS OF FORMER DESIGN

The quartz triangle, which is the fulcrum point on the load side, is very fragile. A considerable amount of precarious repair work is required when it breaks. Translational vibrations of the beam were sometimes observed. These may have been excited by shocks originating from the environment and can have been sustained through time lags in the feedback loop. In combination with nonlinear elements in the loop (e.g., a gradient in the magnetic field in which the coil is immersed) this resulted in dc offset signals in the reading.

Due to these phenomena the a priori linearity of response could not always be confirmed. Deviations up to several μg were found. This does no harm as long as the differences in mass between the sample L and one of the reference weights T_1 or T_2 are small. But for differences exceeding 1 or 2 mg significant errors may appear.

Finally, each single mass determination consists of three readings: T_1, T_2, and L. As the balances are frequently used, a simplification of the reading procedures was considered to be desirable.

MODIFICATIONS

The quartz triangle was replaced by a demountable triangle with an 11-μm tungsten wire. This practically eliminated the risk of breaking this part.

In the torsion suspension at the center of the beam the 50-μm tungsten wires were replaced by stainless steel ribbons of 0.5×0.04 mm cross section. Much higher tension on the suspension limits translational movements of the beam and consequently of the coil in the magnetic field. The total length of the coil was reduced, keeping the number of windings the same. This facilitated the adjustment of the windings in the uniform part of the magnetic field.

The relatively slow and unstable photoresistors were replaced by silicon photodiodes which are much quicker in response and have a much better stability.

As a consequence of the last alterations, and with a view to facilitate the reading of the output signals, the electronic circuit was completely redesigned by the Electronics Department. Figure 2 shows a block diagram.

The differential signal from the photodiodes is fed into a double-stage servoamplifier with an RC filter for critical damping of the beam deflections. The total voltage gain is limited to 200 by feedback loops. The output of the servoamplifier feeds a current through the coil to restore and keep equilibrium of the beam posi-

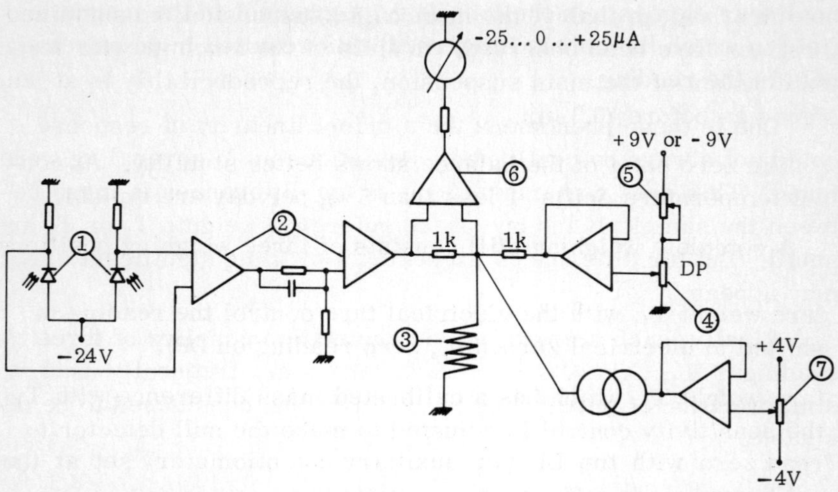

Fig. 2. Electronic circuit. 1) Photodiodes; 2) servoamplifier with damping filter; 3) balance coil; 4) decade potentiometer; 5) sensitivity control; 6) null detector; 7) electrical taring.

tion. The open-loop gain is several thousands (dependent on adjustment of the center of gravity of the beam), as compared with less than 100 with the old servoamplifier. From a decade potentiometer, readable to five figures (DP), an adjustable voltage is fed into a voltage follower. Its output current is proportional to the DP setting. The dial can be adjusted such that the servosystem output current is reduced to zero. Apart from a small offset due to voltage drop across the coil, the output current of the servosystem is then zero, so that practically null-balance is obtained. A potentiometer for sensitivity control provides the possibility to make the DP read directly in μg. Servocurrent zero is detected by an amplifier — galvanometer combination. This offers the following features: amplification of zero signal, signal damping, and overrange protection of the galvanometer. Finally, a current source offers the facility of electrical taring of the balance.

All active components are integrated circuit amplifiers.

RESULTS WITH NEW ELECTRONICS

First results with the new electronics are favorable. Thanks to a higher loop gain, quicker response of the amplifier and photodetectors, and, probably, the more rigid suspension, no sustained vibrations have been observed. In spite of the much greater torsion constant of the main suspension, the reproducibility is at least as good as before (0.3 μg).

The zero point of the balance shows better stability. At constant temperature drifts of less than 5 μg per day are typical.

A complete weighing still consists of three weighing positions:

tare weight T_1; with the electrical tare control the reading is shifted to electrical zero (i.e., zero reading on DP);

tare weight T_2, which has a calibrated mass difference with T_1; the sensitivity control is adjusted to make the null detector to read zero with the DP (or auxiliary potentiometer) set at the calibrated mass difference;

sample; with previous adjustments the DP now reads the mass difference between the sample and the tare weight T_1 directly in μg.

LINEARITY OF RESPONSE

Perfect linearity of response would a priori be expected. An experimental check, however, revealed a number of phenomena which may cause very significant deviations. For example:

a small offset in the zero detector produces the double effect when the difference between negative and positive output signals is to be determined;

swinging of the sample or shocks will result in corresponding oscillations in the feedback current; in combination with nonlinear effects (e.g., saturation of an amplifier) this will produce a net dc offset in the null detection;

linearity of the DP and proportionality of output current of the voltage follower with the DP setting has to be verified; zero reading on the DP might not be zero imput signal, due to contact resistances.

Taking these and similar phenomena into consideration during the design, direct calibration of the balance with standard weights shows no deviations from linearity within experimental error. This error is $\pm 1\ \mu g$, corresponding to the error specified with the standard weights.

ACKNOWLEDGMENTS

The authors acknowledge the constant interest Dr. J. Spaepen, Director of the Bureau, showed in this work. They are, further, greatly indebted to Drs. Verelst and Meyer and Mr. Hubbeling for their valuable and skillful help in the design of the electronic part, and to Messrs. Geerts and Uylenbroek who displayed their usual skill and ingenuity in improving the mechanical parts.

REFERENCES

1. H. Moret and E. Louwerix, Vacuum Microbalance Techniques, Vol. 5 (K. H. Berndt, ed.), Plenum Press, New York (1965), p. 59.
2. A. J. Deruytter, J. Spaepen, and P. Pelfer, Proceedings of the 2nd Conf. on Neutron Cross Sect. and Technology, March 4-7, 1968, Washington, DC.
3. W. van der Eijk and H. Moret, Standardization of Radionuclides, I.A.E.A., Vienna (1967), p. 529.

A New Microbalance Technique for Kinetic Studies of Gas—Metal Reactions at High Temperatures

A. Pebler

Westinghouse Research Laboratories
Pittsburgh, Pennsylvania 15235

ABSTRACT

A microbalance can be used to measure reaction rates of resistance-heated specimens with reactive gases by weighing the accumulation of condensed reaction products on a collector surrounding the hot specimen. The method was used to measure the volatilization rate of tungsten in inert gas in the presence of small amounts of H_2O and CO_2.

INTRODUCTION

Recording microbalances have been widely used to investigate the reactions and vaporization behavior of refractory metals in corrosive environments under static or flow conditions at elevated temperatures.[1–4] With this type of apparatus, the weight change can be followed as a function of time and temperature. Thus kinetic, vapor-pressure, and thermochemical data may be obtained. Usually the sample is suspended from the balance arm in a refractory reaction tube surrounded by a resistance furnace. This arrangement limits the temperature to which the sample can be heated and also introduces unwanted reactions with the hot tube wall which may interfere with the reactions on the sample itself.

In the relatively simple apparatus described here, a metal sample in the form of a wire, ribbon, or filament is electrically heated to incandescence in a stationary position. The sample is surrounded by a bell-shaped fused silica enclosure that is suspended from a recording microbalance and accumulates volatile reaction products formed at the hot sample surface with residual gases at low pressures, or with reactive contaminants in inert-gas atmospheres at normal pressure. Weight changes may be monitored continuously at low pressures or intermittently after appropriate heating intervals at atmospheric pressure. The collector must be at a sufficiently low temperature, i.e., the condensation coefficient must be unity, to condense all the reaction products. Such an arrangement closely simulates conditions in an incandescent lamp and can serve as a means to quantitatively study physicochemical reactions in such lamps. Zavitsanos[5] used a similar arrangement to collect vaporized species effusing from a Knudsen cell.

EXPERIMENTAL

The apparatus shown in Fig. 1 includes the use of a Cahn RG recording microbalance. The sample may be a wire, filament, or ribbon and is attached to heavy inserts of tungsten. The electrical

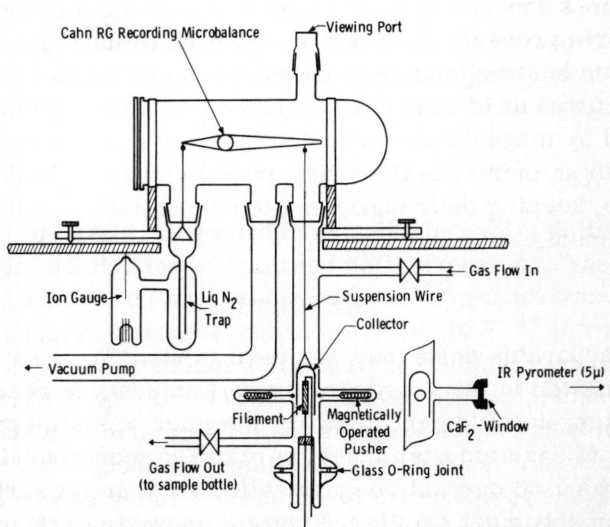

Fig. 1. Microbalance system for high-temperature gas — metal reactions.

feedthroughs are mounted on a separate O-ring glass joint for easy replacement of the sample. The electrically heated sample is surrounded by a bulb which is fabricated from clear fused silica. A loop of fused silica is sealed to the top closed end of the tube in order to suspend the collector from the microbalance. The mass of the collector is reduced by etching to bring it to the capacity of the balance of 1 g. Set screws on the mounting table of the balance case allow one to center the suspended enclosure around the sample.

The wall temperature of the collector is read from an infrared radiation pyrometer operating between 4.8 and 5.6 μ, where fused silica is practically opaque. Thus the temperature reading is not affected by the radiation originating at the incandescent sample. The collector is viewed through a calcium-fluoride window, which transmits over the operating wavelength range of the infrared pyrometer. The sample temperature is measured with an optical micropyrometer prior to the experiment and calibrated in terms of the electrical current, or it may be measured during the experiment by viewing through the bottom opening of the bulb. Transmission and emissivity corrections have to be applied to temperature measurements of either sort.

Weight changes can be monitored continuously if reactions are carried out at low pressures ($<10^{-3}$ torr). The reactive gases and gas mixtures are leaked into the differentially pumped system. At atmospheric pressure the balance is intermittently operated after appropriate heating intervals. Irregular movements of the collector as a result of convection currents in the hangdown tube are prevented by magnetically operated arresters. Reactive gases are leaked into an inert-gas flow and carried slowly through the hangdown tube, keeping their concentration essentially constant during an experiment.

RESULTS

The apparatus described was used in studying the volatilization of tungsten in the presence of small amounts of reacting gases. A coiled tungsten filament of 250 μ diameter was slipped over both the ends of heavy tungsten insert wires. The geometrical heated surface was 1.56 cm^2, although the effective reacting surface area was presumably much smaller. Figure 2 shows an Arrhenius plot of the accumulation rate of reaction products ($WO_2 + W$) on the collector measured in an argon — 10% nitrogen atmosphere with and without a small amount of water vapor added. The weight increase of the collector was intermittently recorded after heating periods which varied between 5 min at high reaction rates and 1 h at very low reaction rates. The duration of the filament heating (5 min minimum) at the same conditions had no definite effect on the deposition rate. Also, a small gas flow in the reaction tube did not change the rate compared to the rate obtained when keeping static conditions in the balance system. It was noted that the deposition occurred mostly at the collector wall next to the filament; this indicates that virtually no convection currents were present in the bell-shaped filament enclosure. Plots of the reaction rates in a wet inert gas (curves 1 and 2, Fig. 2) could be fitted to a straight line. The magnitude of the activation energy of 15.7 ± 1.5 kcal · mole^{-1} indicates that the reaction is transport-controlled under the conditions of this experiment. An empirical heat of activation of 14.3 kcal · mole^{-1} was obtained by Gulbransen et al.[6] in the diffusion-controlled region above 1350 C for the oxidation or pure tungsten in oxygen.

The reaction rate in a pure 10% nitrogen — argon mixture at 760 torr (curve 3, Fig. 2) is about an order of magnitude lower than in the same inert atmosphere containing 0.1% water vapor. The scatter of the data in curve 3 was expected to be greater than

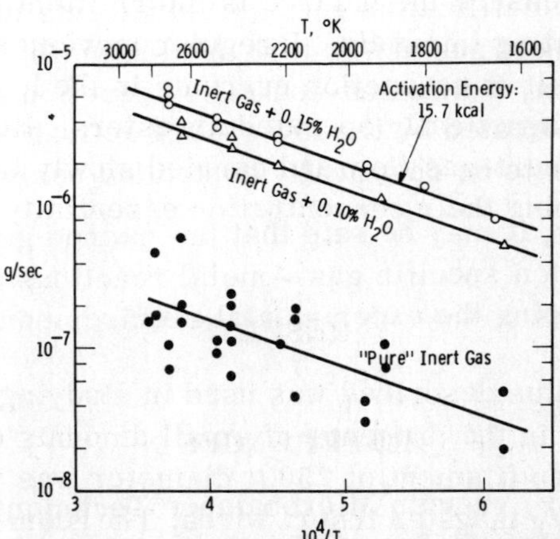

Fig. 2. Mass gain from the volatilization of tungsten in inert gas (90% A + 10% N_2) at 760 torr with and without added water vapor.

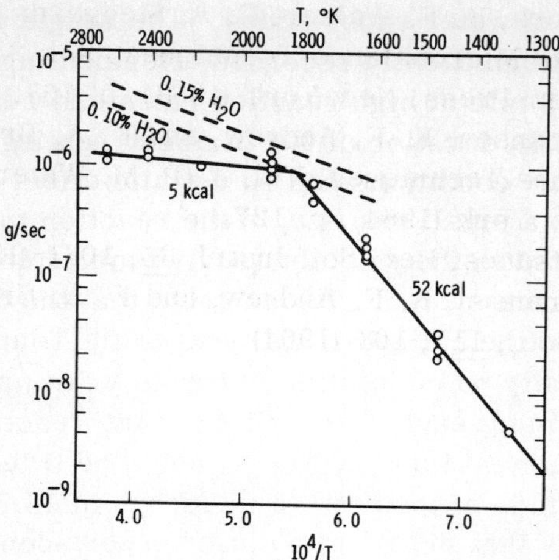

Fig. 3. Mass gain from the volatilization of tungsten in the presence of 0.1% CO_2 in inert gas at 760 torr.

for curves 1 and 2 because of the lower magnitude of the reaction rate and the uncertainties in the actual amount of water vapor present (e.g., the amount of water vapor from surrounding heated glass walls). Nevertheless, the experimental data in curve 3 show about the same temperature dependence as curves 1 and 2.

Results of volatilization rate measurements in the presence of 0.1% CO_2 in inert gas at 760 torr are shown in Fig. 3. At the conditions chosen, a diffusion-controlled reaction persists above about 1800 C, indicated by the low energy activation, whereas the reaction appears to be chemically controlled below 1800 C.

In summary, it may be said that the method gives quick and valuable results on specific gas – metal reactions at high temperatures, while keeping the experimental requirements at a modest level.

REFERENCES

1. R. F. Walker, Vacuum Microbalance Techniques, Vol. 1 (M. J. Katz, ed.), Plenum Press, New York (1961), p. 87.
2. E. A. Gulbransen and K. F. Andrew, Vacuum Microbalance Techniques, Vol. 2 (R. F. Walker, ed.), Plenum Press, New York (1962), p. 129.

3. N. J. Carrera, R. F. Walker, C. A. Steggerda, and W. M. Nalley, Vacuum Microbalance Techniques, Vol. 3 (K. H. Berndt, ed.), Plenum Press, New York (1963), p. 153.
4. E. A. Gulbransen, K. F. Andrew, and F. A. Brassart, Vacuum Microbalance Techniques, Vol. 4 (P. M. Waters, ed.), Plenum Press, New York (1964), p. 127.
5. P. D. Zavitsanos, Rev. Sci. Instr., 35, 1061 (1964).
6. E. A. Gulbransen, K. F. Andrew, and F. A. Brassart, J. Electrochem. Soc., 111, 103 (1964).

A Moving-Table Balance

J. A. Poulis,* M. Verduin,† J. P. de Mey,* and C. H. Massen*

Physics* and Mechanical Engineering† Departments
Eindhoven University of Technology
Eindhoven, Netherlands

ABSTRACT

At the 5th and 6th Conferences of this series the possibilities of using pivot bearings combined with a moving table in balance constructions were indicated. The present paper deals with the use of such a balance incorporating mechanical feedback. The results of measurements concerning the sensitivity (0.12 μg) and reproducibility of such a moving-table balance are discussed.

INTRODUCTION

At the 5th and 6th Vacuum Microbalance Techniques Conferences we reported the research carried out on the applicability of pivot bearings in balance constructions.[1,2] Attention was drawn to the possibility of using pivot-bearings in balances where the effect of a mass variation of the sample is compensated by adjusting the angle of inclination of the supporting plane, as illustrated in Fig. 1. The condition for a horizontal equilibrium position of the balance beam being

$$l\Delta m = mR\sin\Theta_e \tag{1}$$

where

l is the beamlength,

Δm is the mass change of the sample to be determined,

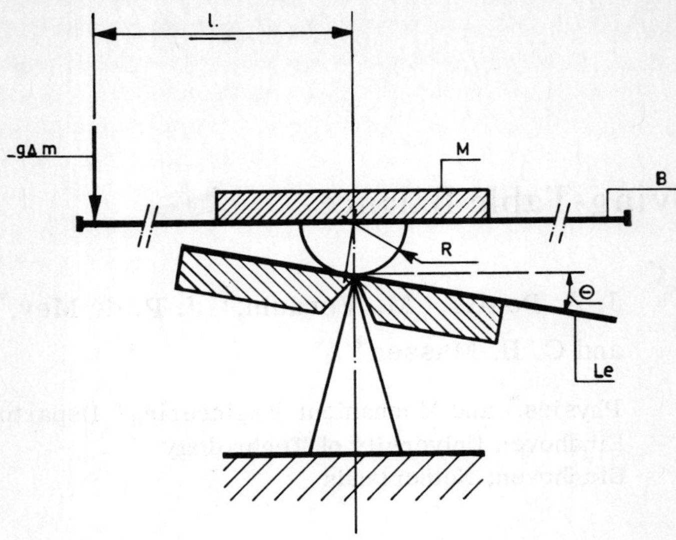

Fig. 1. Schematic diagram of the moving-table balance. For explanation of the symbols, see text.

m is the mass of beam plus counterweight and sample (the mass of the sample not including Δm),

R is the pivot radius, and

Θ_e is the angle of deflection of the table necessary for a horizontal equilibrium of the beam.

This moving-table method for weighing has some favorable features. An advantage it shares with the other balances where the beam is brought back to zero position is that many errors due to inhomogeneities in the environmental conditions (e.g., temperature, magnetic field, etc.) are eliminated. A more specific advantage above the balances, which compensate by means of a Lorentz force, is the absence of ferromagnetic materials and consequently a simpler bakeout procedure for UHV measurements. These considerations led us to construct a moving-table balance and test it on its practical merits. A photograph of this balance was shown at the 6th V.M.T. Conference.[2]

EXPERIMENTAL SETUP

For the tryout measurements of the moving-table balance, use was made of an automatically controlled compensation. A schem-

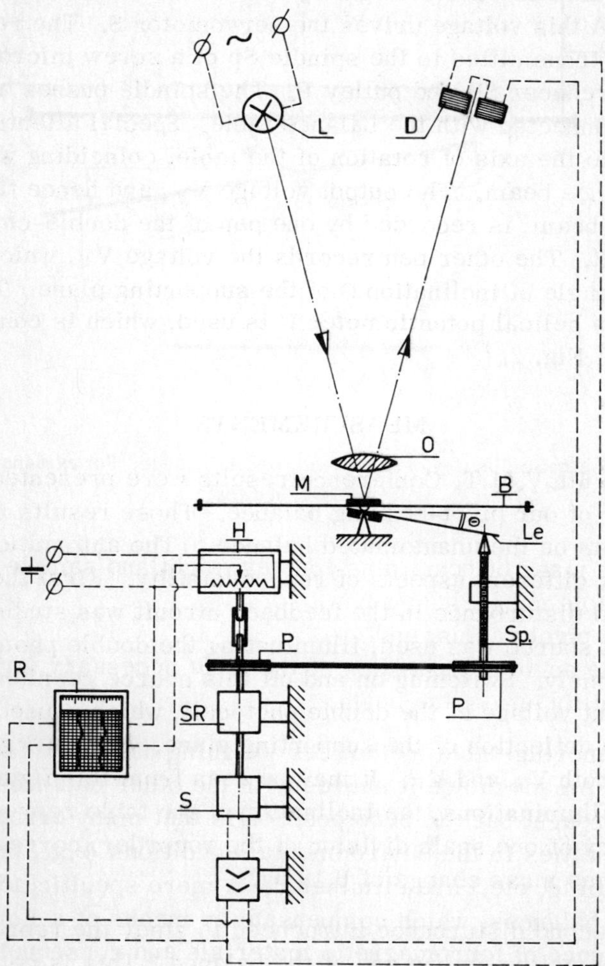

Fig. 2. Schematic feedback system. For explanation of the symbols, see text.

atic of the feedback system is depicted in Fig. 2. The light of a lamp L is reflected by the mirror M attached to the balance beam B. Then it falls on a double photocell D so that a deflection of the balance beam causes a voltage V_B across the cell. Via a servo-amplifier A this voltage drives the servomotor S. The rotation of its axis is transmitted to the spindle Sp of a screw micrometer via speed reducer SR and pulley P. The spindle pushes a lever Le which is connected with the balance table. Special attention has been paid to the axis of rotation of the table, coinciding with that of the balance beam. The output voltage V_B, and hence the deflection of the beam, is recorded by one pen of the double-channel recorder R. The other pen records the voltage V_P, which represents the angle of inclination Θ of the supporting plane. To effect the latter a helical potentiometer H is used, which is connected as depicted in Fig. 2.

MEASUREMENTS

At the 6th V.M.T. Conference results were presented on the sensitivity of our pivot-bearing balance. Those results referred to experiments on the unautomated balance.[2] The automation enables us to study different aspects of reproducibility. First the effect of an external disturbance in the feedback circuit was studied. An extra light source was used, illuminating the double photocell inhomogeneously. Switching on and off this source resulted in an extra output voltage of the double photocell, which caused via the feedback a deflection of the supporting plane. Figure 3 gives the tracks of both V_B and V_P. It may be seen from this figure that, with both illuminations, the inclination of the table reproduces with an accuracy of one scale division of the recorder, corresponding to a spurious mass change of 0.12 μg.

The second disturbance introduced to study the reproducibility was a force acting on one end of the balance. This is comparable to the actual measuring procedure the balance is designed for. The force was produced by bringing a permanent magnet in the vicinity of the beam end. This procedure was repeated several times; each time the distance between the beam end and the magnet was given another value. Consequently, the reproducibility of the zero position after each removal of the extra magnetic force was studied with different values of the extra force. The results of a number

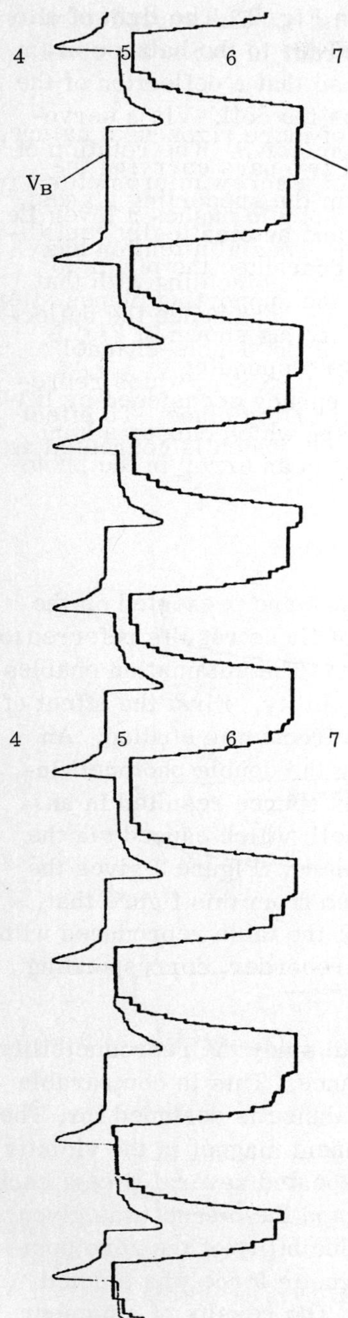

Fig. 3. The influence of an external disturbance in the feedback system on the reproducibility of the balance.

of such experiments may be seen from Fig. 4. The reproducibility of the zero position is apparently of the order of two scale divisions, corresponding to 0.24 μg.

The third disturbance tried out was of more rigorous a nature. The balance beam was lifted by means of two bars carrying the beam, and thus the pivots were lifted from the supporting planes. After that, the beam was put down again and automatically equilibrated. This way of manipulating did not guarantee the pivots to come back at exactly the same points on the supporting planes after each lifting procedure. The reproducibility, as shown in Fig. 5, was not better than 15 scale divisions, corresponding to 1.8 μg. Another possible explanation of the discrepancy occasioned by lifting can be found in nonideal optical imaging which causes a horizontal displacement of the beam to result in an error in the photo-

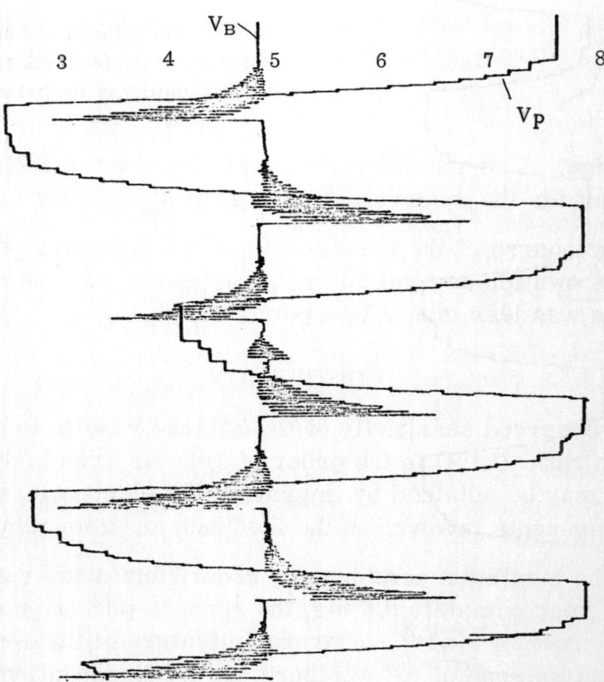

Fig. 4. The influence on the reproducibility of the balance of a force acting on a balance arm.

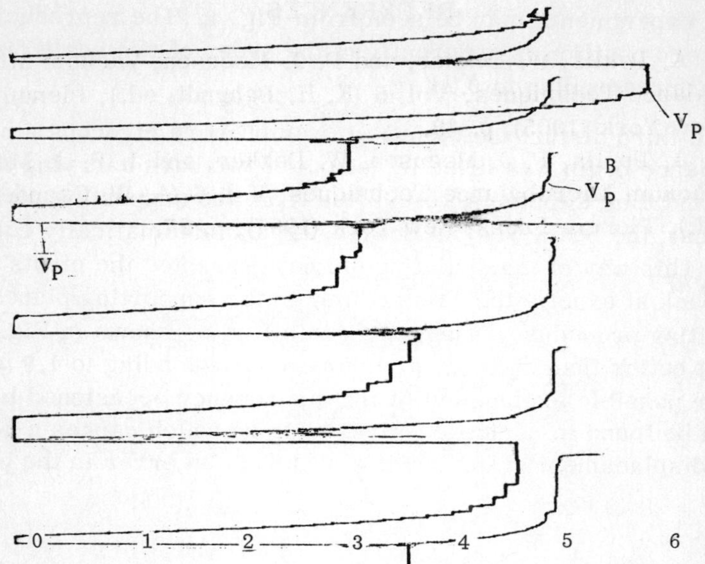

Fig. 5. The influence on the reproducibility of the balance of the lifting of the balance pivots from the supporting plane.

cell voltage. A horizontal-beam displacement of 0.1 mm can readily account for the measured lack of reproducibility.

In the course of the measurements the balance system has been left on its own for several hours. It appeared that the drift of the apparatus was less than 0.1 μg per hour.

DISCUSSION

The observed sensitivity of the balance results in a load-to-precision ratio (LPR) in the order of 10^7. An even higher value of the LPR may be obtained by improving the electronic and mechanical components involved in the feedback and recording circuits.

As the maximum continuously recordable mass variation (span) is approximately 0.6 mg, the span-to-precision ratio (SPR) is of the order of 5×10^3. A favorable feature of the apparatus is that within its span of 0.6 mg there are no direct influences from changes in room temperature, as no magnet for the compensating system is used.

REFERENCES

1. J. A. Poulis, W. Dekker, and P. J. Meeusen, Vacuum Microbalance Techniques, Vol. 5 (K. H. Behrndt, ed.), Plenum Press New York (1965), p. 49.
2. J. A. Poulis, P. J. Meeusen, W. Dekker, and J. P. de Mey, Vacuum Microbalance Techniques, Vol. 6 (A. W. Czanderna, ed.), Plenum Press, New York (1967), p. 27.

Fluctuations of the Weight Indicated by a Microbalance in the Pressure Range Between 1 and 10^3 torr with the Sample at a Lower Temperature than the Beam

E. Robens, G. Sandstede, G. Walter, and G. Wurzbacher

Battelle-Institut e.V.
Frankfurt/Main, W. Germany

ABSTRACT

Large fluctuations in the indicated weight of up to 0.1 Hz frequency and 50 μg amplitude, were observed when using a vacuum microbalance with suspended pans, of which one was cooled to 77 K and the balance beam kept at room temperature. The amplitudes increased with increasing pressure and temperature difference between sample and beam. In a nitrogen atmosphere, the fluctuations began at 200 torr, and in an argon atmosphere at 140 torr, while in helium and hydrogen no fluctuations were observed. When the balance pan was empty, the fluctuations were particularly large, and when loaded, they generally decreased. With horizontally suspended disks and vertical tubes, practically no fluctuations were observed.

A closer investigation of the phenomena showed that the reason for these fluctuations was convection at the pan due to thermal radiation from that part of the balance which was at room temperature. It turned out that the bottom of the pan was acting as a radiation receiver.

Several models are discussed for the interpretation of these observations. The fluctuations were greatly reduced when the pan was shielded with a metal disk attached to the suspension wire immediately above the pan.

INTRODUCTION

In the course of porosity microgravimetric determinations of the capillary condensation of nitrogen at 77 K, continuous fluctuations of the weight reading were observed. The fluctuations, which were noticed at pressures above 200 torr, lead to a substantial reduction of the weight determinations. They could not be attributed to the effects of vibrations in the building, changes in mains voltage, variations in ambient temperature, or gas-pressure fluctuations within the system itself.

It appeared that this effect was due to local convection of the gas at the balance pans. The convection phenomena have been the subject of our investigations.

EXPERIMENTAL ARRANGEMENT

The apparatus "Sartorius Gravimat"[1,2] (manufactured by Sartorius-Werke GmbH, Göttingen, W. Germany) contains a Gast microbalance[3,4] and a pressure gauge which permits the adjustment of pressure to an accuracy greater than 0.1 torr. On the load side of the balance (Fig. 1) a quartz pan 15 mm in diameter and 10 mm in height was suspended from a metal wire about 1 m long and 30 μ thick. This pan was located at about 100 mm above the bottom of the balance tube, which had an inside diameter of 25 mm. A Dewar vessel (internal height 600 mm) filled with liquid nitrogen or oxygen was used for thermostatic control. A wire hook served as a counterweight. The measuring system and counterweight were kept at room temperature. The Dewar vessel was closed with a cover which had aluminum foil glued to its lower side.

The unsymmetrical arrangement of sample and counterweight and the use of various materials caused buoyancy, so that the absolute weight could be determined with an accuracy of only $\pm 20\ \mu g$. The possible presence of an additional constant force due to convection acting on the balance pan[5] was not observable under these conditions; however, measurement of the fluctuations in the weight reading would not be affected by such an error. The fluctuations

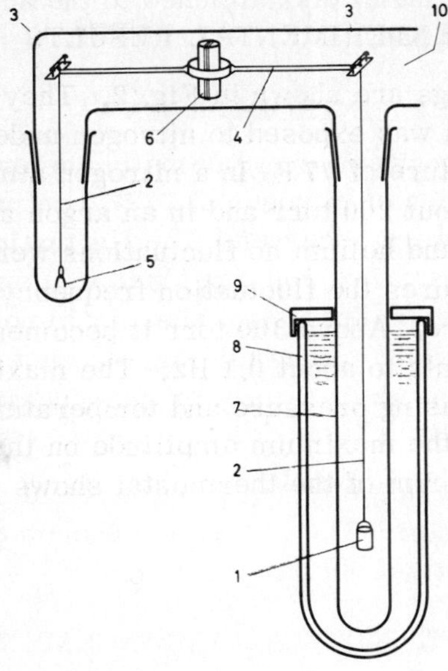
3
3
10
6
4
2
5
9
8
7
2
1

Fig. 1

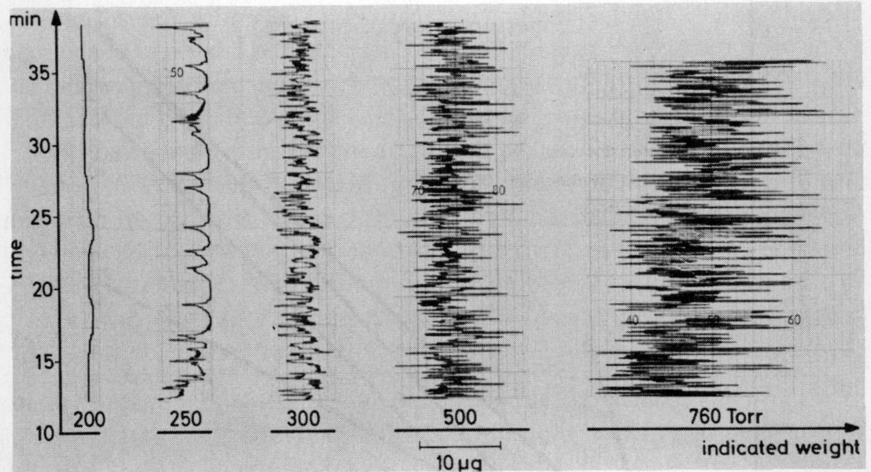
min
35
30
25
20
15
10
time
200
250
300
500
760 Torr
10 μg
indicated weight
50
70
80
40
60

Fig. 2

were observed for at least 30 min, and the error involved in the maximum amplitude indicated was estimated at ±5 μg.

EXPERIMENTAL RESULTS

Typical readings are shown in Fig. 2. They were obtained for an empty pan which was exposed to nitrogen under different pressures at a temperature of 77 K. In a nitrogen atmosphere the fluctuations start at about 200 torr and in an argon atmosphere at 140 torr. In hydrogen and helium no fluctuations were observed. Starting from low pressures the fluctuation frequency increases with an increase in pressure. Above 300 torr it becomes independent of pressure and amounts to about 0.1 Hz. The maximum amplitude increases with increasing pressure and temperature gradient (Fig. 3). The dependence of the maximum amplitude on the distance of the pan from the upper rim of the thermostat shows a maximum as illustrated in Fig. 4.

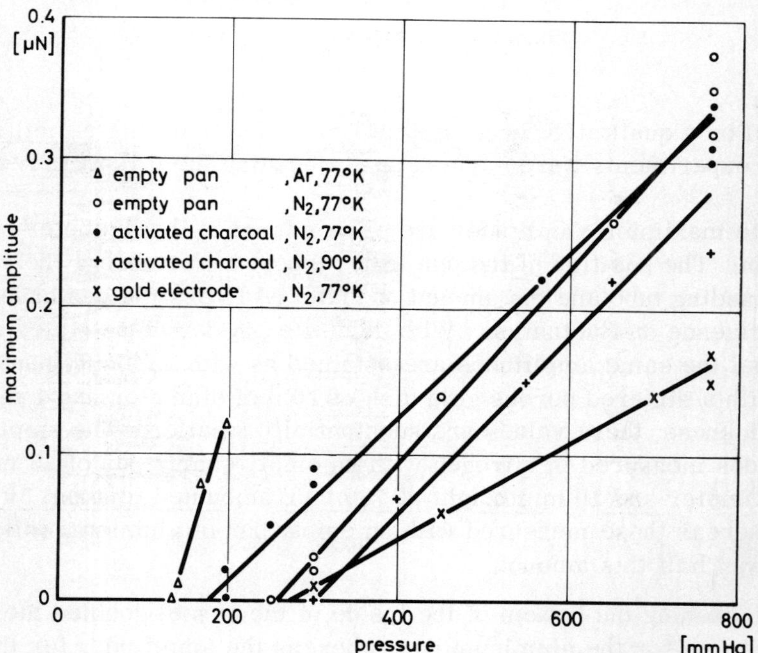

Fig. 3. Fluctuations as a function of pressure and temperature difference.

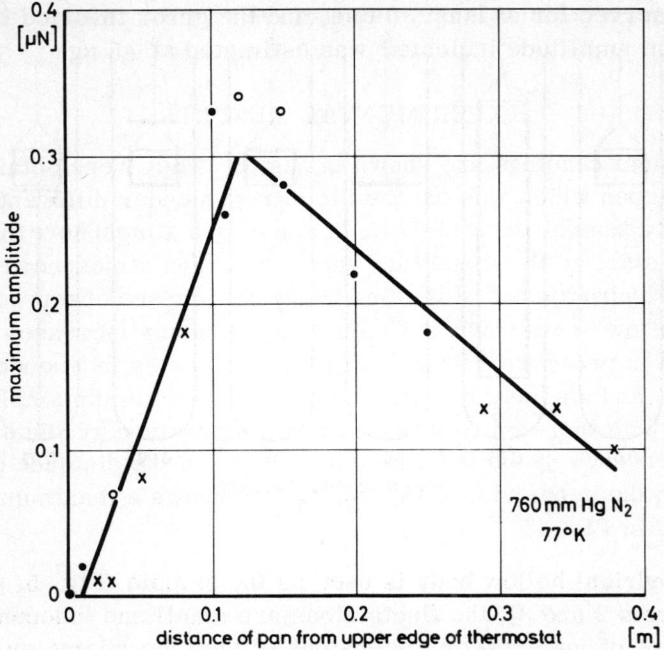

Fig. 4. Fluctuations as a function of position of pan.

For a qualitative investigation of the origin of this effect, several experiments were carried out, producing the following results:

The maximum amplitudes are proportional to the diameter of the pan. The position of the pan with respect to the wall of the surrounding tube and the amount of material filling the pan have no influence on fluctuation. With dolomite powder or activated charcoal the same amplitudes are obtained as with an empty pan. With a sintered porous gold disk of 10 mm diameter and 1 mm thickness, these values are substantially smaller. The amplitudes measured in nitrogen with an empty quartz pan of 15 mm diameter and 10 mm height at 760 torr amounted to about 50 μg, whereas those measured with an empty pan of aluminum foil were about half this amount.

Blackening the bottom of the inside of the vessel doubled the amplitudes for the aluminum pan, whereas the amplitudes for the quartz pan remained unchanged.

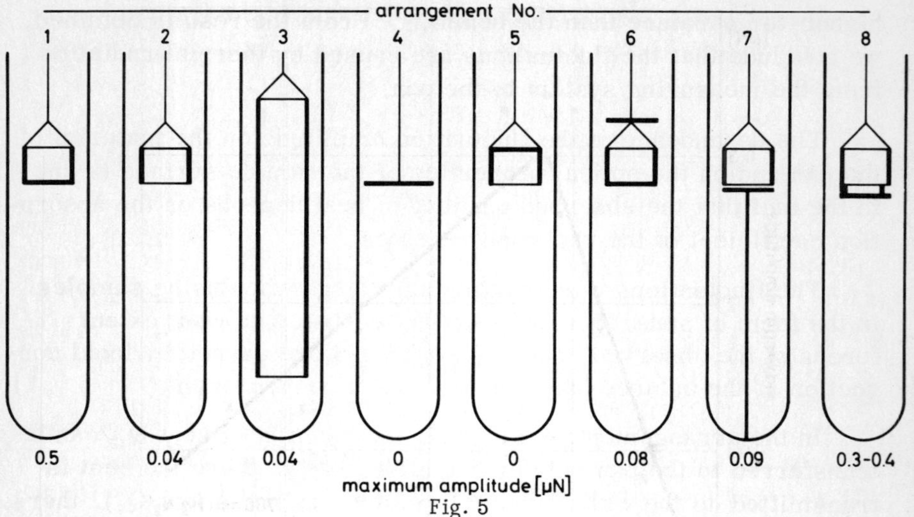

Fig. 5

If a cylindrical hollow body is used as the sample (Fig. 5, arrangements 2 and 3), the fluctuations are small and independent of the type of material or dimensions of the tube (diameter 4 to 15 mm, length 10 to 400 mm). The maximum amplitude is not larger than 4 μg even in the range of large temperature gradients (about 2 deg $\cdot$ mm^{-1}).

A disk alone (arrangement 4) or an aluminum pan turned upside down (arrangement 5) show practically no fluctuations. If the pan is shielded with a metal disk (arrangement 6), the maximum fluctuation amplitudes are reduced to 8 μg. The fluctuations increase with the increasing distance of the disk from the pan.

If an aluminum disk, fixed to the balance tube is placed under a suspended glass tube of 15 mm diameter and 10 mm height (arrangement 7), minor fluctuations occur which increase with decreasing distance of the disk from the suspended glass tube. In this case, the maximum amplitude at a distance of 1 mm is 9 μg. If the disk is attached to the glass tube (arrangement 8), the fluctuations are as large as 35 μg at a distance of 2 mm and 40 μg if the disk is tightly glued to the glass tube.

DISCUSSION

The fluctuations of the weight readings agree in their profile with those observed upon normal convection (with the sample at a higher temperature than the beam).[6,7] From the results obtained, we conclude that the fluctuations are caused by thermal radiation from the measuring system to the pan.

The dependence of the fluctuation amplitude on the material of the pan and on the optical properties of the sample surface is due to the fact that the absorbed quantity of heat depends on the absorption coefficient of the radiation receiver.

The fluctuations were particularly large when using samples in the form of pans. It will be estimated below to what extent forces of the observed order of magnitude are caused by local convection at the balance pan when using this arrangement.

In the arrangement considered, the quantity of heat Q_{12} can be transferred to the pan only by radiation. From there the heat is transmitted to the wall of the balance tube by radiation (Q_r), thermal conduction, and convection (Q_c). Thus, under steady-state conditions,

$$\dot{Q}_{12} = \dot{Q}_r + \dot{Q}_c \tag{1}$$

$\dot{Q}_{12}$ can be calculated[8] if we approximate the arrangement by two parallel plates A_1 and A_2, spaced at a distance r_1:

$$\dot{Q}_{12} = \frac{\varepsilon_1 \varepsilon_2 \sigma A_1 A_2}{\pi r_1^2} T_1^4 \tag{2}$$

For T_1 = temperature of the weighing system = 298 K
A_1 = cross-sectional area of the balance tube = $5 \times 10^{-4}\,\mathrm{m}^2$
A_2 = cross-sectional area of the bottom of the pan = $2 \times 10^{-4}\,\mathrm{m}^2$
r_1 = distance of the pan from the upper edge of the thermostat 0.1 m
ε = emissivity; $\varepsilon_1 = 1$; $\varepsilon_2 = 0.9$
σ = Stefan — Boltzmann constant = 5.6×10^{-8} watt · m^{-2} · deg^{-4}

we obtain a heat flux of $\dot{Q}_{12} = 1.3 \times 10^{-3}$ watt toward the pan.

For the radiation exchange between concentric surfaces used in approximating radiation emitted from the pan, we may write[8]:

$$\dot{Q}_r = \left\{ \sigma A_2' / [(1/\varepsilon_2 + A_2'/A_3)(1/\varepsilon_3 - 1)] \right\} (T_2^4 - T_3^4) \tag{3}$$

where A_2' = surface area of the pan = $7 \times 10^{-4}\,\mathrm{m}^2$

A_3 = surface area of the balance tube involved in the heat exchange = $10^{-2}\ m^2$
T_2 = temperature of the pan
T_3 = temperature of the balance tube = 77 K
ε_3 = emissivity of the balance tube = 0.9.

In vacuum, heat is exchanged only by radiation. In this case the heat balance (1) is reduced to

$$\dot{Q}_{12} = \dot{Q}_r \tag{1a}$$

Using Eqs. (2) and (3), the steady-state temperature of the pan in vacuum is calculated to be T_2 = 92 K, which agrees well with the results of direct measurements in an almost identical arrangement.

If gas is present in the balance tube, the term $\dot{Q}_c$ in Eq. (1) has to be taken into account. The lower limit for the $\dot{Q}_c$ term can be estimated by the following equation developed for conduction in convection-free gas in a spherically symmetrical system:

$$\dot{Q}_c = \left[4\pi\lambda/(1/r_2 - 1/r_3)\right](T_2 - T_3) \tag{4}$$

where r_2 = radius of the pan = 7.5×10^{-3} m
r_3 = radius of the balance tube = 1.4×10^{-2} m
λ = thermal conductivity[9];
at 77 K in N_2, $\lambda_{N_2} = 6.7 \times 10^{-3}$ watt · deg^{-1} · m^{-1};
in H_2 and He, $\lambda_{N_e} = 5.4 \times 10^{-2}$ watt · deg^{-1} · m^{-1}
T_3 = 77 K

By inserting Eqs. (2), (3), and (4) into (1), the upper limit of the pan temperature is estimated to be 78.5 K in nitrogen and 77.1 K in hydrogen and helium atmospheres. As heat exchange between pan and liquid-nitrogen bath by conduction is for a factor of about 10 larger than by radiation ($\dot{Q}_c \approx 10\dot{Q}_r$), we will confine our discussion to the last term in Eq. (1).

Our experiments have indicated that the gas is not free of convection. Hence:

(1) the conduction—convection term $\dot{Q}_c$ becomes larger than would be predicted by Eq. (4), so the temperature difference between pan and balance tube is reduced;

(2) the moving gas acts mechanically on the balance pan.

An upper limit of the mechanical work L per unit time, which may occur in the form of kinetic energy of the convecting gas, may be estimated by assuming reversible transport of the irradiated heat $\dot{Q}_{12}$ from the pan to the nitrogen bath by means of a Carnot cycle:

$$L = \dot{Q}_{12}(T_2 - T_3)/T_3 \tag{5}$$

In nitrogen atmosphere this power amounts to 1.6×10^{-5} watt, in hydrogen and helium to 2.2×10^{-6} watt.

Furthermore, it is possible to estimate the maximum work done on the balance from the observed oscillations. If we assume the oscillation of the balance to be harmonic, and thus the force to be proportional to the deflection, we obtain the average power $\overline{L}$ delivered to the balance as:

$$\overline{L} = (1/\pi)k_0 x_0^2 \omega \tag{6}$$

where k_0 = force constant
x_0 = amplitude of pan oscillation
ω = angular frequency.

Assuming also a maximum deflection x_0 of $0.2\,\mu$ then $\overline{L}$ would be 6×10^{-13} watt. This amount of power may be even smaller for quasielastic behavior. The oscillatory power picked up by the balance is much smaller than the power available by irradiation, and would be sufficient to cause the described fluctuations. In order to interpret the more significant experimental observations, i.e.,

(a) maximum weight fluctuation amplitude,

(b) mean frequency of fluctuations,

(c) dependence of fluctuation amplitudes on the nature of ambient gas atmosphere,

we considered two models of momentum transfer:

1. The gas inside the pan expands due to an increase in its temperature, a gas bubble breaks off and generates recoil momentum on the balance.

2. Eddy currents are formed in the gas within the pan and generate a force on the balance by momentum transfer to the wall of the pan.

Both models agree only partially with experimental observations. The calculated values of fluctuation frequency are in both cases considerably greater than those observed. Model (1) yields qualitatively a correct dependence of fluctuation amplitude on the molecular weight of the gas (approximately $\propto M^{\frac{1}{2}}$), but the maximum amplitude would then be too high ($14\,\mu N \approx 1400\,\mu g$). Model (2) gives, in nitrogen atmosphere, the correct fluctuation amplitude but too weak a dependence on the molecular weight of the gas (approximately $\propto M^{1/4}$).

On the whole, we feel that the experiments show that the fluctuations are due to convection caused by the absorption of very small amounts of radiation energy by the balance pans. The detailed pattern of gas flow and the mechanism of momentum transfer from the moving gas to the balance is, however, not yet clear and should perhaps be studied in greater detail.

In order to suppress the fluctuations, a flat shielding plate should be attached to the suspension wire immediately above the pan. An additional reduction of the influence of thermal radiation may be achieved by choosing materials of high reflectivity for the shield plate, e.g., bright metals. This method of shielding has already been recommended[10] for thermostatic control in vacuum. In this case the thermal radiation increases the temperature of the shield from 77 K to about 78 K, so that the equilibrium temperature of the pan exceeds the temperature of liquid nitrogen by not more than a few millidegrees.

ACKNOWLEDGMENTS

For valuable discussions, our thanks are due to Dr. H. Büttner, Dr. E. Gerlach, and Dr. D. Langbein, Battelle-Institut e.V., Frankfurt/Main.

REFERENCES

1. E. Robens and G. Sandstede, Z. Instrumentenk., 75, 167-178 (1967).

2. E. Robens and G. Sandstede, J. Sci. Instr., Series 2, 2, 365-368 (1969).
3. Th. Gast, Z. Angew. Phys., 8, 164-171 (1956).
4. B. Kassner, Chem. Rundschau, No. 18, pp. 1-4 (1959).
5. J. D. Ferchak, Rev. Sci. Instr., 38, 273-275 (1967).
6. C. Cahn and H. Schultz, Analyt. Chem., 35, 1729-1731 (1963).
7. W. Kuhn, E. Robens, G. Sandstede, and G. Walter, this volume, pp. 161-172.
8. See, for example, E. Schmidt, Einführung in die technische Thermodynamik, 9th Edition, Springer, Berlin (1962), pp. 350, 404, 405.
9. See, for example, G. K. White, Experimental Techniques in Low Temperature Physics, Oxford University Press, London (1959), pp. 71, 187, 190.
10. G. Sandstede and E. Robens, Chem.-Ing.-Techn., 32, 413-417 (1969).

An Improved, Highly Sensitive, and Bakeable Microbalance System with a Built-in Calibration Device for Studying Condensation Phenomena Between —128 and 70 C in UHV

P. Schmider and H. Mayer

Department of Physics
Technische Universität Clausthal
Clausthal, W. Germany

ABSTRACT

Two microbalances, I and II, commonly support a sample pan at the (lower) end of a quartz fiber. System I is an electrostatically compensated balance with a tungsten-wire suspension. System II is mounted at right angles below system I, and has no weighing function.

This arrangement offers several advantages in UHV microbalance techniques. First, sideward pendulum oscillations of the long pan suspension are eliminated. Secondly, this design offers the possibility of using the torsion wires of microbalances I and II as leads for electrical and thermoelectrical measurements on the material deposited on the sample pan.

For the first time it was possible to readjust and calibrate our microbalance system in situ under UHV conditions by stepwise loading the pan with weights ranging from 3 up to 500 μg. Measurements of the temperature-dependent condensation behavior of a Cd-atom beam on quartz pan and simultaneous measurements of electrical film resistance were performed.

MEASURING CELL

For the experimental investigation of heterogeneous condensation of metallic atoms on solid dielectric surfaces a microbalance system was developed which yielded several improvements with respect to those described earlier.[1] The experiments, of which some results are given below, were performed in a measuring cell, including the microbalance system. Figure 1 shows the main construction features of this cell.

In the lower part of the cell there was a Cd-atom beam source from which emerged a constant beam of Cd atoms (see No. 1 in Fig. 1). After entering the cooled part of the cell through suitable slits, the intensity of the Cd-atom beam could be measured by a 6-MHz quartz oscillator (see No. 5 in Fig. 1), which was moved magnetically. Through further slits the atom beam reached the middle part of the cell, which consisted of a double-walled highly reflecting glass Dewar (see No. 6 in Fig. 1) and which could be filled with liquefied gases (e.g., O_2, N_2, or H_2) by means of a connecting tube.

In the center of the Dewar and perpendicular to the direction of the impinging flux, a thin quartz sheet (see No. 7 in Fig. 1) of dimensions $20 \times 20 \times 0.2$ mm was provided as a substrate for the condensation process. A heating coil near the quartz pan was used for the adjustment of higher temperatures in order to reevaporate the condensate. The pan was suspended from the microbalance system by a thin quartz fiber (length 35 cm).

In Fig. 2 the balance system and the sample pan are to be seen in more detail. The temperature of the pan was measured in the following way. On the back of the pan was pressed a second quartz sheet around which was wound a 10-μ Pt resistance wire. After calibration of this arrangement in a separate vacuum system, it showed a good reproducibility and a temperature sensitivity of 3 $\Omega \cdot \text{deg}^{-1}$. In order to obtain the resistance of the growing metal film, two narrow Pt contacts were evaporated on opposite sides of the pan.

BALANCE SYSTEM

In order to get reliable results on problems of heterogeneous condensation the microbalance system had to meet the following

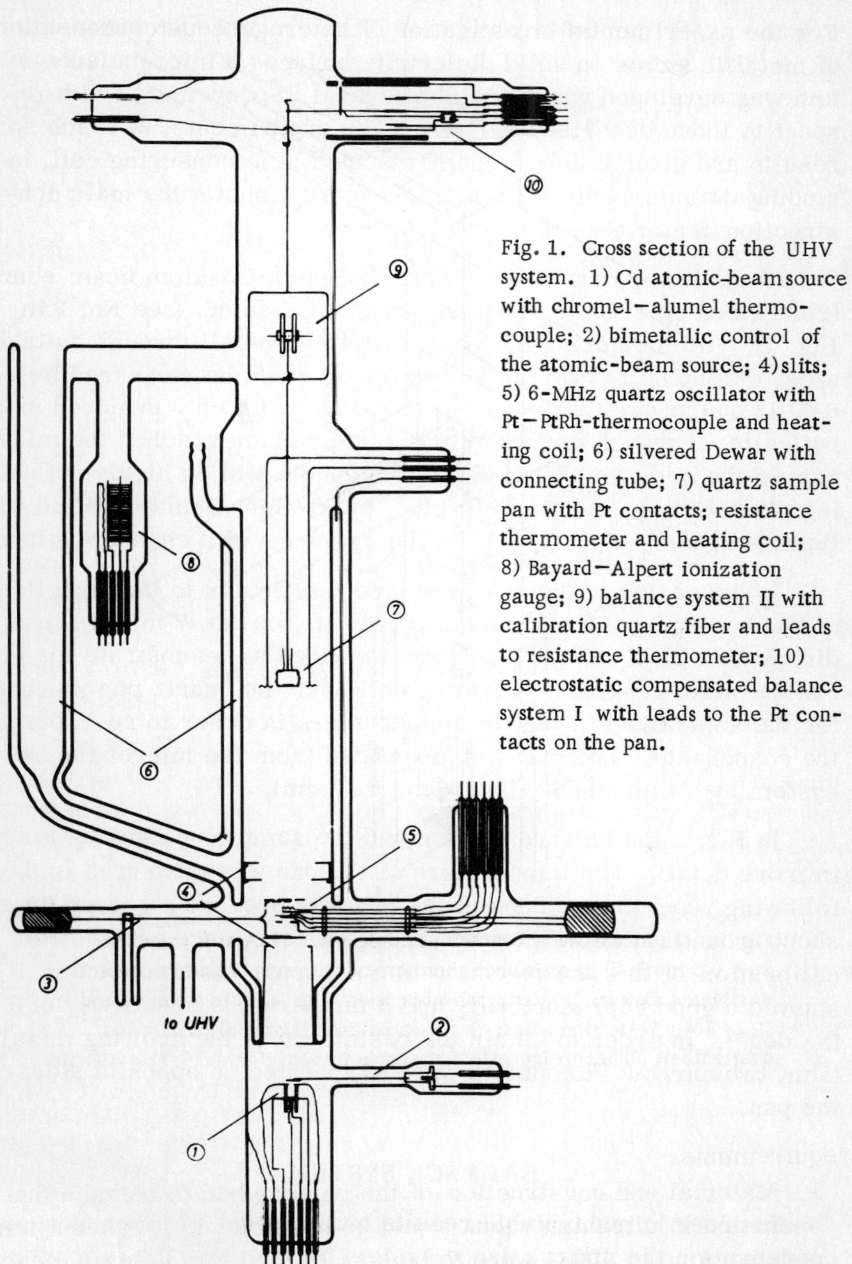

Fig. 1. Cross section of the UHV system. 1) Cd atomic-beam source with chromel–alumel thermocouple; 2) bimetallic control of the atomic-beam source; 4) slits; 5) 6-MHz quartz oscillator with Pt–PtRh-thermocouple and heating coil; 6) silvered Dewar with connecting tube; 7) quartz sample pan with Pt contacts; resistance thermometer and heating coil; 8) Bayard–Alpert ionization gauge; 9) balance system II with calibration quartz fiber and leads to resistance thermometer; 10) electrostatic compensated balance system I with leads to the Pt contacts on the pan.

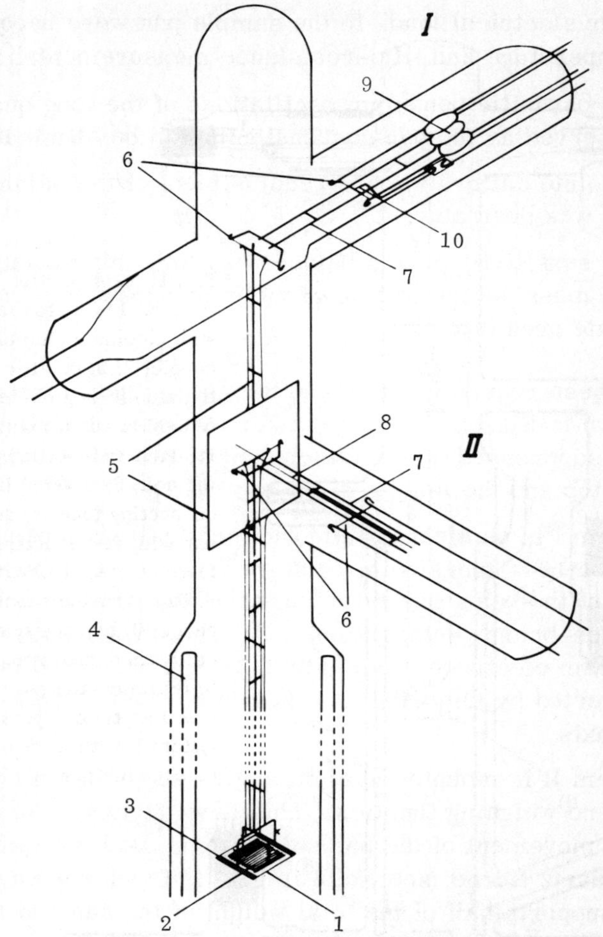

Fig. 2. Microbalance system with sample pan. 1) Quartz pan (20 × 20 × 0.2 mm); 2) Pt resistance thermometer; 3) evaporated Pt contacts; 4) silvered Dewar; 5) quartz pan suspension with leads to system I and II; 6) 10-μ W-torsion wires; 7) quartz balance beam; 8) calibration quartz fiber; 9) beam electrode in the capacitance bridge; 10) 7-μ Au wire.

requirements:

1. Material and construction of the balance had to be such that operation in ultrahigh vacua would be possible, i.e., bakeout temperatures up to 400 C were to be applied in order to obtain clean and gas-free surfaces.

2. Four electrical leads to the sample pan were necessary for the temperature and film-resistance measurements.

3. The parasitic pendulum oscillations of the long quartz fiber which served as the pan suspension had to be eliminated.

4. Absolute calibration and readjustment of the balance system in UHV was desirable.

5. The sensitivity of this balance had to be high enough to permit the mass determination of parts of monatomic layers on the substrate area of 4 cm^2.

All these requirements could be met by the microbalance device shown in Fig. 2. The sideward movement of the pan suspension was suppressed by the attachment of two microbalances, I and II, to the top and the middle of the quartz fiber.

System I is an electrostatic, automatically compensated balance with a 10-μ tungsten-wire suspension.[2,3] Compensation of deflections of the beam caused by loads on the pan was achieved by a capacitance-bridge detector method. The compensation voltage for the beam electrode was fed by a 7-μ Au wire. To minimize the torque exerted by this wire, it was wound several times around the fulcrum axis.

System II is mounted at right angles about 15 cm below system I and has no weighing function. This arrangement stabilizes the sideward movement of the pan suspension. Both systems are joined by a quartz frame (see No. 6 in Fig. 2) in such a way that each balance supports half of the total weight of the pan and the pan suspension (about 600 mg). The beams of the balances consisted of two quartz fibers isolated from each other, so that the torsion wires of, e.g., system I, could be used as leads for the film resistance measurements on the pan and those of system II as leads for the resistance thermometer on the back of the sample pan. The resistance of the tungsten wires — about 300 Ω — is an additional contribution; the leads along the pan suspension consisted of 50-μ Ag wires.

The metallic junctions of wires and the mechanical attachment of wires to the quartz were done by means of "Einbrennsilber." *

* "Einbrennsilber" 108 supplied by "Degussa."

CALIBRATION OF THE BALANCE

The main advantage of this balance design was the additional possibility for calibration and readjustment. For this purpose, in system II a narrow quartz tube was fused to the beam. Through this tube a thin quartz fiber (length, 80 mm; diameter, 82 μ; mass, m = 1.06 mg) could be moved (see No. 8 in Fig. 2). A shift Δx of the quartz fiber corresponded to a weight change ΔM at the pan of

$$\Delta M = m(\Delta x / l) \quad (1)$$

where l is the length of the lever arm (for system II, l = 34.5 mm). The movement of the quartz fiber was achieved by means of bi-

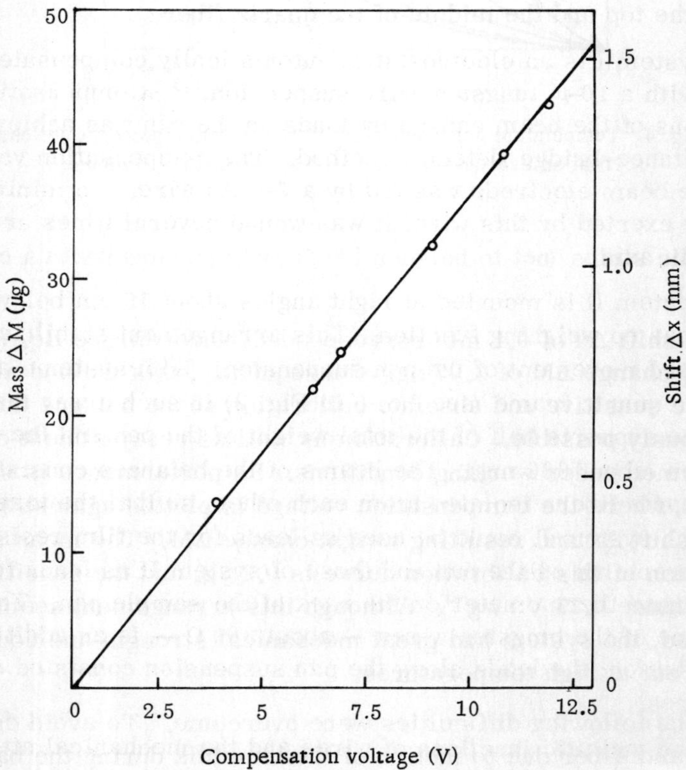

Fig. 3. Quartz-fiber shift Δx and mass loading ΔM versus compensation voltage V.

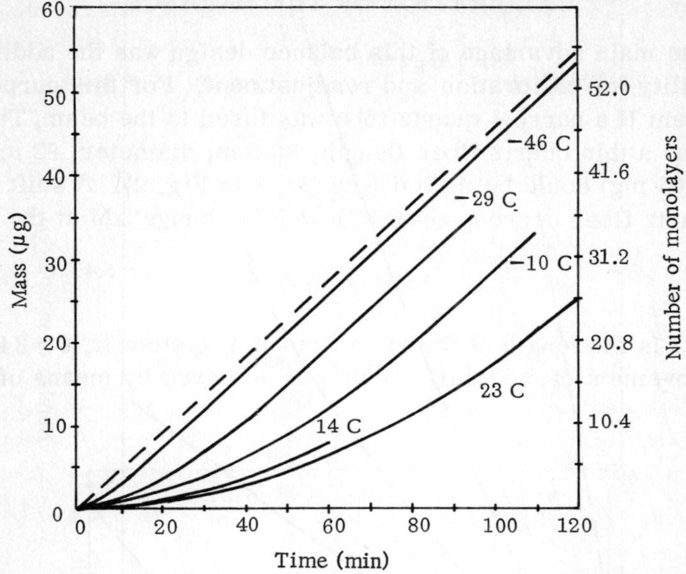

Fig. 4. Temperature dependence of the condensation process Cd — quartz (I) at an evaporation rate of 0.47 monolayer per minute.

metallic strips (not to be seen here) and measured with a cathetometer.

A shift Δx of 0.1 mm corresponded, according to Eq. (1), to a weight change ΔM of 3.07 μg. Consequently, this method allowed a more sensitive and absolute calibration of the balance than was previously possible. Calibration as well as readjustment could be performed under working conditions. Its performance is shown in Fig. 3, where the compensation voltage is plotted against quartz fiber shift Δx and resulting weight change ΔM. The mean square deviation of this calibration curve is 0.2 μg, and the sensitivity of the balance 0.25 V $\cdot$ μg^{-1}. Although this arrangement seems complicated, the system had great mechanical strength and could be baked out at high temperatures.

The following difficulties were overcome. To avoid damaging the quartz fiber due to motion of the bimetals during the bakeout process, the bimetals were constructed with a "thick" and a "thin" part. Under the influence of an electric current only the "thin" part moved, causing a shift Δx, but when all parts were warmed up

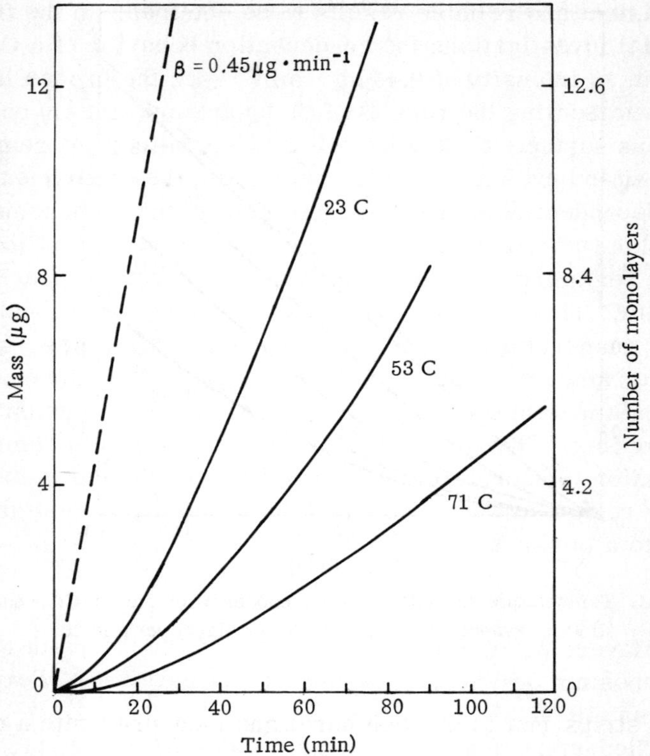

Fig. 5. Temperature dependence of the condensation process Cd – quartz (II) at an evaporation rate of 0.45 monolayer per minute.

simultaneously at bakeout temperatures, the "thick" part compensated for the movement of the "thin" one. In order to avoid spurious charges on the glass surfaces, the cell was made conducting by means of SnO.

CONCLUSIONS AND RESULTS

In conclusion, it can be said that this balance design, which was developed with the aim of performing satisfactory condensation experiments, is a mass detector in UHV with a sensitivity of 0.2 monolayers Cd on an area of 4 cm^2; it allows direct and exact measurements of the temperature and the film resistance on the sample pan. But the main advantage of the arrangement is that the in situ calibration before and after each measurement permitted consistent and reliable results to be obtained. In the first experimental investigations the condensation behavior of a Cd-atom beam with an intensity of 0.45 $\mu g \cdot min^{-1}$ (see the broken line in Fig. 4 representing the results of the quartz oscillator) on the amorphous surface of vitreous silica was studied. At room temperatures, and even at lower temperatures, the experiments showed a time-dependent condensation coefficient α and a phenomenon called "delayed condensation."[4] Complete condensation (α = 1) from the beginning is brought about at a substrate temperature near −50C. This corresponds to a supersaturation of about 10^9. At elevated temperatures condensation also occurred (see Fig. 5), but in smaller amounts at higher temperatures. At 78 C the evaporation rate of bulk material on the pan is the same as the intensity of the impinging flux. There is no hint of the existence of a "critical" condensation temperature. All Cd films investigated in the temperature region between −128 and 70 C showed a resistance $\geq 10^{12}$ ohm up to a thickness of 280Å.

REFERENCES

1. H. Mayer, W. Schroen, and D. Stünkel, 1960 Seventh National Symposium on Vacuum Technology Transactions, Pergamon Press (1962).
2. R. Niedermayer and W. Schroen, Vakuumtechnik, 11, 36 (1962).
3. D. Hillecke and R. Niedermayer, Vakuumtechnik, 3, 69 (1965).
4. H. Göhre, Physik und Technik von Sorptions- und Desorptionsvorgängen bei niederen Drücken; Vortrag des 2. Europäischen Symposium "Vakuum" (1963).

[illegible] beam with an intensity of 0.45 [illegible] broken line [illegible] the [illegible] surface [illegible] and even at lower temperatures, the experiments showed [illegible]

called "delayed condensation." Complete [illegible]

[illegible] This corresponds to a [illegible] of about [illegible] At [illegible] smaller amounts at higher temperatures [illegible] the evaporation [illegible] the intensity of the [illegible] condensation temperature [illegible] have been investigated. In the temperature [illegible] up to a thickness of 250 Å.

REFERENCES

1. R. Mayer, W. Schoenes and D. Stuart, 1960 Seventh National Symposium on Vacuum Technology Transactions, Pergamon Press (1960).
2. R. Niedermayer and [illegible] Schroeder, [illegible] 16 (1960).

[illegible] material. It is shown [illegible] in addition to experiments on these [illegible] has been made of the [illegible] crystal powders of chemical compounds.

Results which are complementary to the [illegible] measurements can be obtained with [illegible] quantitative de- [illegible] mass spectrometric analysis. [illegible] the course of de- [illegible] partial pressure at [illegible]

Experimental Results and Theoretical Considerations on Thermogravimetric Decomposition Reactions of Chemical Compounds Under High Vacuum

H. G. Wiedemann

Mettler Research Laboratories
Greifensee, Switzerland

ABSTRACT

Investigations on thermogravimetric decomposition reactions under high vacuum have demonstrated that special attention should be paid to such factors as the sample size, the crucible form, the thickness of the layer in the crucible, and the particle size of the probe material. It is shown how the various influences can be eliminated by observing certain starting or test conditions. In addition to experiments on these specific influences, a comparison has been made of the degradation speeds of single crystals and crystal powders of chemical compounds.

Results which are complementary to the thermogravimetric measurements can be obtained with qualitative or quantitative determination of the gaseous decomposition products by mass-spectrometric analysis. From these measurements the course of decomposition was followed by recording the partial pressure at the various individual masses. This method is particularly suitable for the study of thermal degradations which produce more than one decomposition product in the same step.

INTRODUCTION

The results in the field of thermal analysis show that thermogravimetric investigations become even more important when they can be carried out under vacuum.[2] The field of application ranges from simple thermal decomposition to the measurement of the kinetics of chemical reactions, including the conditioning of samples before carrying out reactions in a special gas atmosphere (e.g., the removal of adsorbed layers such as water or CO_2 which could interfere with the process under examination). Especially in the thermogravimetric decomposition of a chemical compound the sample size is dependent on the quantity of the resulting gaseous decomposition products. An idea of the correct sample size can generally be obtained from the performance of the pumping system which is combined with the balance. A short pressure increase during such measurements cannot be avoided. It must not, however, exceed the admissible pressure of the diffusion pumps. This means that samples for thermal decomposition in high vacuum should be limited to a few mg. So as to avoid secondary reactions, in particular sorption effects during decomposition reactions, the layer of the substance in the crucible must be kept as thin as possible. This requires, even for small samples, plate-type crucibles. The layer thickness of the substance should not exceed 0.5 to 1 mm.[13]

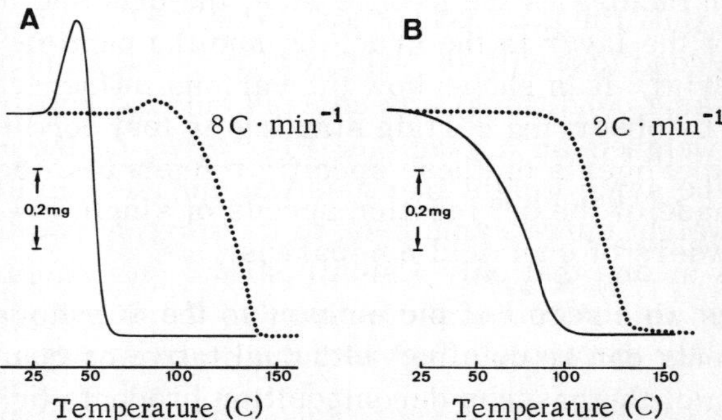

Fig. 1. Dehydration of Whewellite ($CaC_2O_4 \cdot H_2O$). Solid line, powdered material (particle size 150 mesh); dotted line, single crystal (ca. 1 × 1 × 2 mm); all samples weigh 6 mg. A) Heating rate, 8 C · min^{-1}; B) heating rate, 2 C · min^{-1}.

Only in the case of thin layers is a steady warm-up possible because, in the range of 25 to 650C the heat transmission takes place mainly via the crucible. The selection of the heating rate is in many cases dependent on the purpose of an investigation; it is recommended, however, to use small heating rates (see Figs. 1A, B).

For studying the kinetics of a decomposition, probes with different particle size were used. Figure 1 (A and B) shows the dehydration of Whewellite powder ($CaC_2O_4 \cdot H_2O$, 150 mesh) and of a single crystal ($1 \times 1 \times 2$ mm) of Whewellite under identical conditions: a sample mass of 6 mg, heating rates of 2 and 8 $C \cdot min^{-1}$, and an initial pressure of 10^{-5} torr. In this example the pronounced difference in the reactions shows the significance of the particle size and the heating rate.

ERRORS AND EFFECTS IN VACUUM THERMOGRAVIMETRY

Several types of errors, including buoyancy effects, radiometric effects, and thermomolecular flow, can influence vacuum thermogravimetry. Some of these have been discussed in other volumes of this series.[9-11] One effect which occurs during the thermal decomposition in vacuum will be discussed here.

Figure 2A shows as an example the decomposition of calcium oxalate. During heating an increase in weight up to a certain maximum can at first be seen, after which the actual weight decrease takes place (full line). So as to be able to define this effect more precisely, a second test with the Whewellite (same sample size and under the same conditions) was carried out; the substance, however, was weighed on a ring (see Fig. 3) beside the empty crucible, and the same weight increase was measured; after decomposition the weight curve came back to its starting position (dotted line). This means that only a small part of this effect ($<0.5\%$) can be attributed to a recoil of the molecules leaving the balance-pan, but practically the whole effect is due to a reimpact.

Figure 3A shows the furnace tube with a plate-type crucible and ring, and Figs. 3B and 3C show, as described above, the two kinds of experimental procedures with the different arrangements of the probe. Figure 2A shows the result of the two measurements as well as the curve (broken line) obtained by sub-

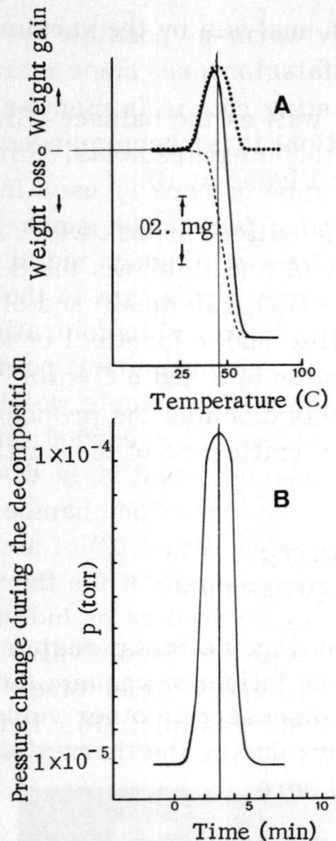

Fig. 2. Dehydration of Whewellite ($CaC_2O_4 \cdot H_2O$). A. Solid line, weight change during the reaction taking place on the balance-pan (plate-type crucible); dotted line, weight change during the reaction taking place beside the balance-pan (ring); broken line, curve diagrammatically determined out of the measured curves. B. Pressure change simultaneously measured during dehydration of calcium oxalate.

traction of the two recorded curves, which serves as the basis for kinetic measurements.[6] The simultaneously recorded pressure change (Fig. 2B) was the same for both decompositions. The same effect was observed by Houde[5] when he vaporized silver on a balance pan; because vaporization took place from below against a suspended balance pan, a weight decrease was simulated. Friedman[12] also has seen this effect in the decomposition of polymers.

MEASURING EQUIPMENT

In order to investigate the observed effect of weight gain at the beginning of a decomposition under variable conditions, the process of decomposition may be simulated on a vacuum thermobalance.[3,4] This is done by introducing gases in measured quantities through a separate inlet into the reaction chamber above the balance pan. The balance housing is evacuated before each analysis by the vacuum system to be approximately 5×10^{-6} torr.

The reaction chamber (Fig. 4D), as well as the balance (G), is kept at 25 C by a thermostat during the measurements. For the measurements up to 1000 C a quartz tube furnace is used instead of the temperature-regulated furnace tube. The gases which are used for the investigation are directed through metal supply pipes to the needle valve (C) with servomotor, and to the outlet (R). The gas flow is continuously measured by an ionization gauge and recorded on the recorder (H). The measuring point (S) is arranged next to the reaction chamber. The sample weight, the preliminary vacuum, and the temperature are recorded on the recorder (O). The temperature-measuring point is in direct contact with the plate-type crucible in the reaction chamber (D). Above one of the diffusion pumps (K) an analyzer (F) of a mass spectrometer[8] is fitted into the vacuum system of the thermobalance. Combined with the recorder (J) mass spectra or individual masses can be recorded continuously. With the needle valve (M) different pressures can be regulated by sucking in air or other gases through the gas purification (N) and the gas inlet (P). Before

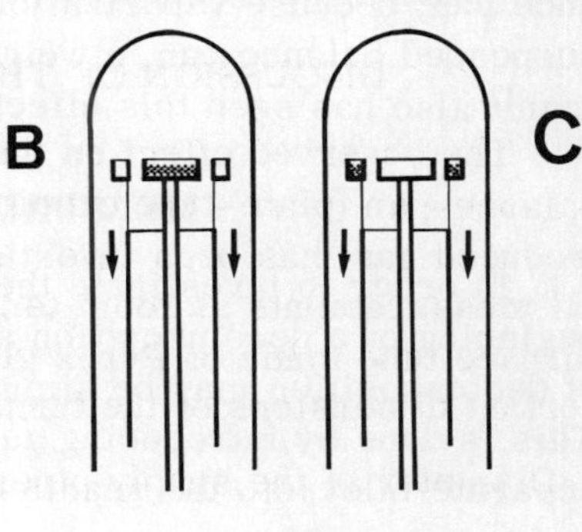

Fig. 3. A) Furnace tube with plate-type crucible and ring; B) plate-type crucible with substance arranged on the balance; C) ring with substance beside the empty plate-type crucible.

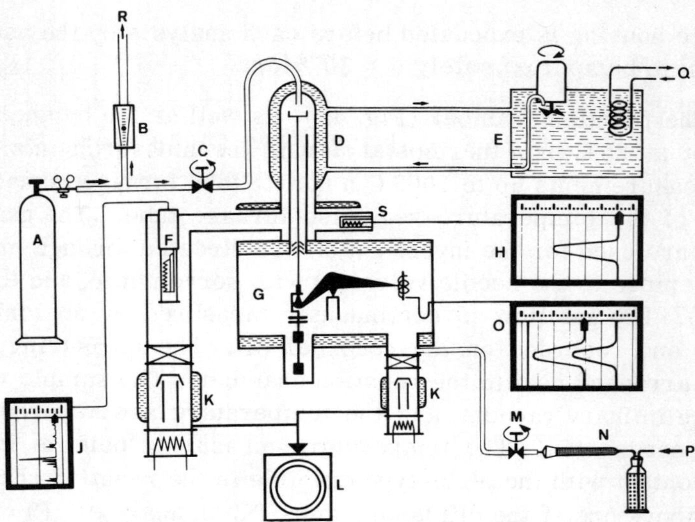

Fig. 4. Experimental equipment. A) Gas cylinder; B) flowmeter; C) needle valve with servomotor; D) thermostatically controlled reaction chamber; E) thermostat; F) analyzer of mass spectrometer; G) thermostatically controlled balance compartment and balance; H) recorder for pressure registration (high vacuum); J) recorder (mass spectrometer); K) diffusion pumps; L) rotary pump; M) needle valve; N) gas-purifier; O) 12-channel multipoint recorder (recording of the weight, temperature, preliminary vacuum); P) intake stack for air; Q) thermostat cooler; R) gas outlet; S) high-vacuum measuring cell.

each measurement the starting pressure of 1×10^{-5} torr was adjusted by this valve. The weight changes were recorded in a balance sensitivity range corresponding to 1000 μg per full chart width. Figure 5A shows an original diagram with an indication of the sensitivity per scale division.

DISCUSSION OF THE MEASURED RESULTS

The observed effect on the apparent increase of the mass of a balance-pan (plate-type crucible), when charged by impulse with introduced gas, has been investigated under different conditions. For all measurements at room temperature, a temperature-regulated furnace tube made of Pyrex glass was used (Fig. 6). The most important dimensions of the furnace tube are:

Diameter of the supply pipe: 3 mm

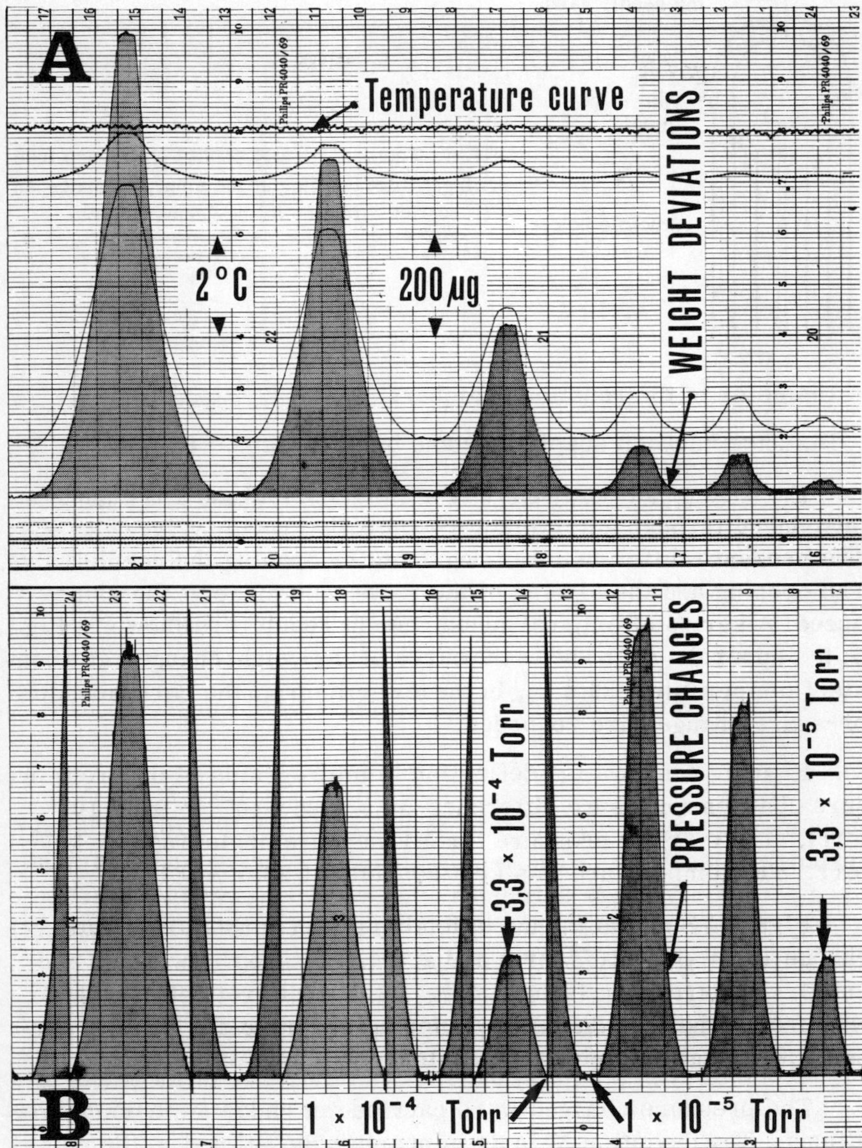

Fig. 5. Diagram (A) shows different weight changes which are due to the introduction of different quantities of argon into the reaction chamber above the pan of the balance evacuated to 1 × 10^{-5} torr. The corresponding pressure changes which have been recorded simultaneously with weight changes can be seen in the diagram (B).

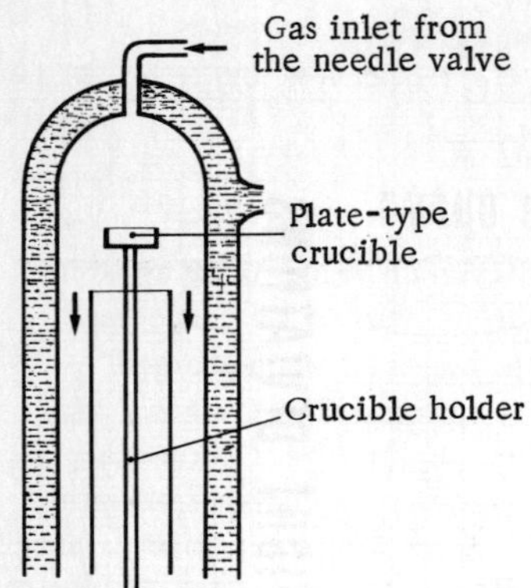

Fig. 6. Thermostatically controlled furnace tube with gas inlet pointing at the empty balance-pan → , suction direction of the vacuum system.

Diameter of the reaction chamber: 45 mm

Distance between supply pipe and balance-pan: ~85 mm

Purified argon was used as the experimental gas. The gas was introduced through a servomotor-controlled needle valve (Fig. 4C). The pressure difference was regulated by opening the needle valve. The opening and the closing of the valve took place at the same speed. A series of original measurement results are shown in Fig. 5; the upper diagram shows the curve of the weight changes at a constant temperature (25.5 C). In the lower diagram the corresponding pressure changes are recorded.

Figure 7A shows the deflection in micrograms at different pressure changes (pressure difference $\Delta p = 1.25 \times 10^{-5}$ to 9.2×10^{-4} torr). The starting pressure for all measurements was 1×10^{-5} torr. The temperature was kept constant at 25.5 ± 0.1 C.

Figure 7B shows the deviation in μg at a constant pressure change ($\Delta p = 8.2 \times 10^{-5}$ torr) and at temperatures between 25 and 985 C. A quartz-tube furnace (Fig. 4B, C) with reflector system was used.[3] the introduced gas being heated up to the furnace temperature before entering the reaction chamber.

The pressure changes were carried out under isothermal conditions. The deviation M in μg is nearly proportional to the root of the absolute temperature. With the aid of a regression analysis the following relation may be obtained:

$$M\ (\mu g) = aT^{\frac{1}{2}} + b \cdot T \qquad (1)$$

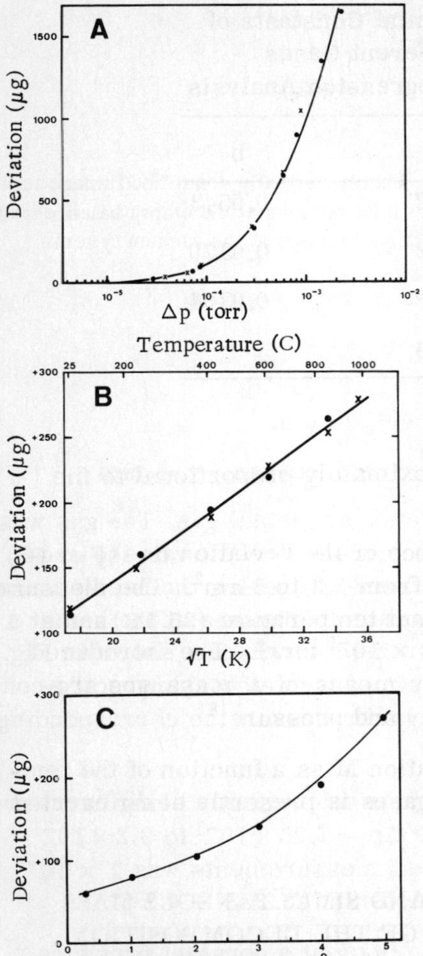

Fig. 7. Maximum values of an apparent increase of the mass of a balance pan (plate-type crucible) when charged by impulse with introduced argon. Diameter of the supply pipe, 3 mm; diameter of the reaction chamber, 45 mm; distance between supply pipe and balance pan (plate-type crucible), 85 mm; starting pressure in reaction chamber, 1×10^{-5} torr. A. Constant temperature, 25.5 C; impact surface of balance pan (plate-type crucible), 1.16 cm^2; change of the pressure difference Δp from 1.25×10^{-5} to 9.2×10^{-4} torr (Δp by different valve openings). B. Temperature isotherm in steps between 25 and 985 C. Impact surface of balance-pan (plate-type crucible), 1.16 cm^2; constant pressure change Δp, 8.2×10^{-5} torr (Δp by equal valve openings); temperature measured by thermocouples of Pt – PtRh (10%) directly on the plate-type crucible. C. Constant temperature, 25.5 C; impact surface of balance-pan (plate-type crucible) from 0.3 cm^2 to 5 cm^2; constant pressure change Δp, 8.2×10^{-5} torr (Δp by equal valve openings).

These measurements have also been carried out on carbon dioxide, nitrogen, and oxygen. The same kind of temperature dependence was obtained, but with slightly different constants. The constants of the different gases are summarized in Table I.

The correction values b are very small. This would allow measurement of a high gas temperature in relation to the apparent mass increase. According to the Maxwell theory, this appar-

Table I. Experimental Constants of Eq. (1) for Different Gases as Obtained by a Regression Analysis

Gas	a	b
Ar	5.92	0.0543
N_2	6.32	0.0370
O_2	6.78	0.0304
CO_2	5.29	–

ent mass increase should be approximately proportional to the average velocity of the molecules.

Figure 7C shows the dependence of the deviation on the crucible surface areas, which varied from 0.3 to 5 cm^2. The measurements were carried out at a constant temperature (25.5C) and at a constant pressure change ($p = 8.2 \times 10^{-5}$ torr). The introduced gas was continuously analyzed, by means of a mass spectrometer, as a further check on purity and pressure.[8]

Further research on the deviation M as a function of the temperature up to 2000 C for various gases is presently being carried out.

THERMAL DECOMPOSITION AND SIMULTANEOUS MASS SPECTROMETRIC ANALYSIS OF THE DECOMPOSITION PRODUCTS[6]

The Balzers Quadrupol high-frequency mass spectrometer Q101[8] which was used consists of two units — the analyzer and the electric control unit. In our case, the mass analyzer was installed directly into the vacuum system of the thermoanalyzer.[3,4]

Simultaneous mass-spectrometric investigations, besides TG and DTA analysis[7] in high vacuum (already mentioned), require a certain sample size not to be exceeded. The amount of gas released during decomposition has also to be compatible with the pumping speed of the vacuum system. Furthermore, the maximum pressure for the mass spectrometer should not be exceeded.

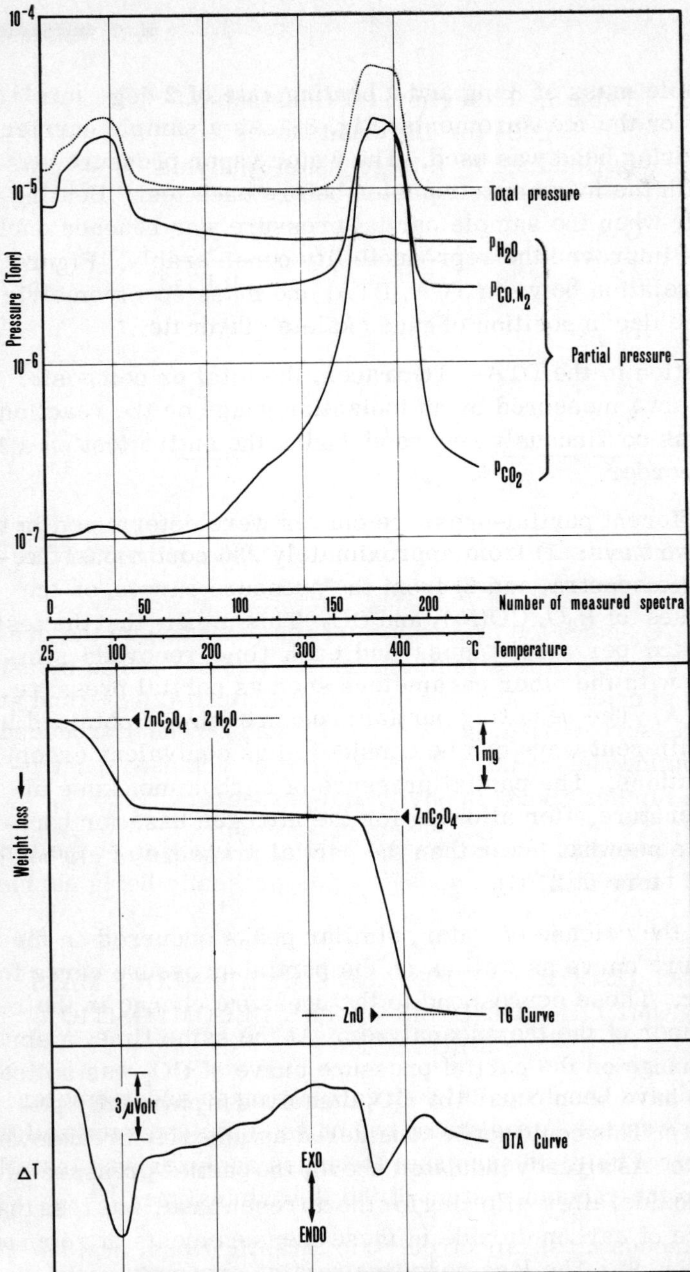

Fig. 8. Decomposition of $ZnC_2O_4 \cdot 2H_2O$ in high vacuum combined with a mass-spectrometric analysis of effluent gas. Total-pressure curves as well as pressure curves for the separate masses of water (18), CO/N_2 (28), and CO_2 (44) are shown next to the simultaneously recorded TG and DTA traces as functions of temperature.

A sample mass of 8 mg and a heating rate of 2 deg · min^{-1} were used for the measurements (Fig. 8). As a sample carrier a DTA measuring head was used. The water vapor pressure was checked with the mass spectrometer before each test. Heating was started only when the sample partial pressure was reached each time. This improved the reproducibility considerably. Figure 8 shows the relation between TGA, DTA, and mass spectrometric results for the decomposition of zinc oxalate dihydrate.

In addition to the DTA — TG traces, the total or composite pressure curve measured by an ionization gauge on the reaction chamber was continuously recorded during the entire test on a 12-channel recorder.

The different partial-pressure curves were determined in the following two ways: 1) from approximately 230 continuously recorded single spectra, and 2) from the pressure curves of the single masses of H_2O, $CO(N_2)$, and CO_2. This means that the tests were repeated per single mass and each time recorded simultaneously with the other parameters such as partial pressure, TG, and DTA. The resulting partial pressure curves obtained in these two different ways can be considered as equivalent except for minor deviations. The partial pressure of carbon monoxide at room temperature after allowing for the nitrogen base (or background), is somewhat lower than the partial pressure of carbon dioxide ($\sim 10^{-7}$ torr at 25 C).

During the release of water, similar peaks occurred on the total-pressure curve as well as on the partial-pressure curve for water vapor. These correspond to the pressure change in the reaction chamber of the thermoanalyzer. At the same time, a small pressure change on the partial pressure curve of CO_2 was noticed, which could have been caused by CO_2 dissolved in physically absorbed water. This could not be considered as an early breakdown of the oxalate. As already indicated above, the partial pressure of carbon monoxide, after allowing for the nitrogen base, was less than the pressure of carbon dioxide in these measurements at room temperature (Fig. 8). The less sensitive partial-pressure curve (ten times) seems to indicate that the beginning of the CO release takes place at a considerably lower temperature than the CO_2 release, which begins at $\sim$150 C. In principle, the CO release is similar with respect to temperature to the CO_2 release. When the TG and pressure curves are compared, it shows that increasing CO_2 release is about at the same height as the end of the water release. A small amount of water can be detected during the CO/CO_2 release. The DTA curve shows a clearly defined peak for the main water release, which is followed by a wavy line indicating a reverse reaction on the dehydrated oxalate with the water. The CO/CO_2 release shows up as an additional DTA peak. The DTA trace approaches the base line again only after complete release of the carbon dioxide.

ACKNOWLEDGMENTS

The assistance of Mr. A. van Tets in theoretical questions, and of Mr. H. Meier in translation, is gratefully acknowledged.

REFERENCES

1. R. Giovanoli and H. G. Wiedemann, Helv. Chim. Acta, 50, 127 (1968).
2. M. Harmelin and C. Duval, Microchim. Acta, 1, 17 (1967).
3. H. P. Vaughan and H. G. Wiedemann, Vacuum Microbalance Techniques, Vol. 4 (P. H. Waters, ed.), Plenum Press, New York (1965).
4. H. G. Wiedemann, Chem. Ing. Technik, 36, 1105 (1964).
5. A. L. Houde, Vacuum Microbalance Techniques, Vol. 3 (K. H. Behrndt, ed.), Plenum Press, New York (1963).
6. H. G. Wiedemann, A. van Tets, and H. P. Vaughan, paper presented at the Pittsburgh Conference on Analytical Chemistry and Applied Spectroscopy, February 21, 1966.
7. R. S. Golke and H. G. Langer, Anal. Chem., 37, 433 (1965).
8. Balzers AG, Balzers, Principality of Liechtenstein, Quadrupol high-frequency mass spectrometer, QMG 101, Bulletin P 51-30e.
9. Vacuum Microbalance Techniques, Vol. 1 (M. J. Katz, ed.), Plenum Press, New York (1962).
10. Vacuum Microbalance Techniques, Vol. 2 (R. F. Walker, ed.), Plenum Press, New York (1962).
11. Vacuum Microbalance Techniques, Vol. 3 (K. H. Behrndt, ed.), Plenum Press, New York (1963).
12. H. L. Friedman, Anal. Chem., 37, 768 (1965).
13. H. G. Wiedemann, Thermal Analysis, Vol. 1, p. 229, Academic Press, New York (1969).

Author Index

Underscored numbers indicate chapters in this volume.

Subject Index